Game Development with Rust and WebAssembly

Learn how to run Rust on the web while building a game

Eric Smith

BIRMINGHAM—MUMBAI

Game Development with Rust and WebAssembly

Copyright © 2022 Packt Publishing

Publishing Product Manager: Rohit Rajkumar

Senior Editor: Aamir Ahmed

Content Development Editor: Feza Shaikh

Technical Editor: Simran Udasi

Copy Editor: Safis Editing

Project Coordinator: Manthan Patel

Proofreader: Safis Editing

Indexer: Manju Arasan

Production Designer: Prashant Ghare

Marketing Coordinators: Elizabeth Varghese and Teny Thomas

First published: April 2022

Production reference: 1310322

Published by Packt Publishing Ltd.

Livery Place

35 Livery Street

Birmingham

B3 2PB, UK.

ISBN 978-1-80107-097-3

www.packt.com

To my wife, Crystina, who believes in me even when I don't, and our kids, Anthony, Niko, Sebastain, Leilani, and Quinn.

— Eric Smith

Contributors

About the author

Eric Smith is a software crafter with over 20 years of software development experience. Since 2005, he's worked at 8th Light, where he consults for companies big and small by delivering software, mentoring developers, and coaching teams. He's a frequent speaker at conferences speaking on topics such as educating developers and test-driven development, and holds a master's degree in video game development from DePaul University. Eric wrote much of the code for this book live on his Twitch stream at `www.twitch.tv/paytonrules`. When he's not at the computer, you can find Eric running obstacle races and traveling with his family.

I want to thank my employer, 8th Light, for allowing me a little time at the end of this process to wrap up this book. I also want to thank my friends in the #rust Slack channel for their help on thornier problems, and the Rustaceans team on Twitch for supporting me and this book.

About the reviewers

Brooks Patton has over 19 years of experience in the IT world, including teaching and game development. He has been programming using Rust for about four of those years as of the time of writing.

His experience with Rust includes game development using the **GGEZ** (**Good Game Easy**) framework, web servers using Actix Web and Rocket, and WebAssembly for non-game web applications.

Brooks has authored a free course on YouTube on creating an **ECS** (**Entity Component System**) – an alternate way of organizing state and functions to mutate that state to the one shown in this book – in Rust using a TDD system.

Joseph McCormick is a software consultant with 8th Light with a background in both e-commerce and internal platform services. He currently specializes in frontend web application architecture and interface development, has an enthusiastic fondness for Rust, and is excited about the promise of a Rust-filled future.

Table of Contents

Part 2: Writing Your Endless Runner

3
Creating a Game Loop

4
Managing Animations with State Machines

5
Collision Detection

Part 3: Testing and Advanced Tricks

9
Testing, Debugging, and Performance

10
Continuous Deployment

11

Further Resources and What's Next?

Preface

The Rust programming language has held the "most-loved" technology ranking on Stack Overflow for 6 years running, while JavaScript has been the most-used programming language for 9 years straight, as it runs on every web browser (`https://bit.ly/3JBg4ms`). Now, thanks to WebAssembly (or Wasm), you can use the language you love on a platform that's everywhere. This book is an easy-to-follow reference to help you develop your own games, teaching you all about game development and how to create an endless runner from scratch. You'll begin by drawing simple graphics in the browser window, and then learn how to move the main character across the screen. You'll also create a game loop, a renderer, and more, all written entirely in Rust. After getting simple shapes onto the screen, you'll scale the challenge by adding sprites, sounds, and user input. As you advance, you'll discover how to implement a procedurally generated world and add sound effects and music. Finally, you'll learn how to keep your Rust code clean and organized so that you can continue to implement new features and deploy your app on the web. By the end of this Rust programming book, you'll have built a 2D game in Rust, deployed it to the web, and be confident enough to start building your own games.

Who this book is for

This game development book is for developers interested in Rust who want to create and deploy 2D games to the web. Game developers looking to build a game on a web platform using WebAssembly without C++ programming or web developers who want to explore WebAssembly along with JavaScript web will also find this book useful. The book will also help Rust developers who want to move from the server side to the client side by familiarizing them with the WebAssembly toolchain. Some knowledge of Rust programming is assumed, but you do not need to be an expert.

What this book covers

Chapter 1, Hello WebAssembly, sets up your first WebAssembly project, explains the toolchain, and runs an application in the browser, drawing to the HTML Canvas that we'll be using throughout this book.

Chapter 2, Drawing Sprites, introduces you to our main character, Red Hat Boy, by showing you how to render a .png file to the screen. Then, we'll make Red Hat Boy run with animation and a sprite sheet.

Chapter 3, Creating a Game Loop, introduces a very basic game engine, so that we can move our character all around the screen at 60 frames per second.

Chapter 4, Managing Animations with State Machines, describes how to make Red Hat Boy run, slide, and jump with state machines and the Rust typestate pattern.

Chapter 5, Collision Detection, starts to make the game fun, making Red Hat Boy crash into and jump over obstacles. We'll introduce axis-aligned bounding boxes and tweak them to account for transparency.

Chapter 6, Creating an Endless Runner, takes the game from one scene into a scene where Red Hat Boy runs to the right, jumping procedurally generated obstacles and platforms that continue for as long as you can keep playing.

Chapter 7, Sound Effects and Music, shows us how to use the Web Audio API to get real immersion into the game with sound effects and catchy music.

Chapter 8, Adding a UI, integrates HTML with the canvas to create a UI, restructuring the game to make it fit.

Chapter 9, Testing, Debugging, and Performance, helps us write some automated tests for the game and investigates performance with the browser tools.

Chapter 10, Continuous Deployment, deploys our game to the web so that anybody can play!

Chapter 11, Further Resources and What's Next?, takes us through what to do next for bigger, more ambitious games.

To get the most out of this book

This book expects you to have a rudimentary understanding of Rust and doesn't cover the syntax. It doesn't expect you to be an expert; you won't be writing any macros or complicated traits, so even a Rust cheat sheet should be enough. The book is structured as a tutorial and is best worked through beginning to end.

Software/hardware covered in the book	Operating system requirements
`rustup`	Windows, macOS, or Linux.
`wasm-pack`	Windows, macOS, or Linux.
`webpack`	Windows, macOS, or Linux.
HTML Canvas	Any modern web browser. This code has been tested on Firefox, Chrome, and Brave.
Netlify CLI	Windows, macOS, or Linux.
GitHub Actions	Windows, macOS, or Linux.
TexturePacker	Windows, macOS, or Linux.

The code for this book was tested with Rust version 1.57.0. Most of the tools will be installed automatically by `rust-webpack-template`.

If you are using the digital version of this book, we advise you to type the code yourself or access the code from the book's GitHub repository (a link is available in the next section). Doing so will help you avoid any potential errors related to the copying and pasting of code.

Download the example code files

You can download the example code files for this book from GitHub at `https://github.com/PacktPublishing/Game-Development-with-Rust-and-WebAssembly`. If there's an update to the code, it will be updated in the GitHub repository.

We also have other code bundles from our rich catalog of books and videos available at `https://github.com/PacktPublishing/`. Check them out!

Code in Action

The *Code in Action* videos for this book can be viewed at `https://bit.ly/3uxXl4W`.

Download the color images

We also provide a PDF file that has color images of the screenshots and diagrams used in this book. You can download it here: `https://static.packt-cdn.com/downloads/9781801070973_ColorImages.pdf`.

Conventions used

There are a number of text conventions used throughout this book.

`Code in text`: This indicates code words in text, database table names, folder names, filenames, file extensions, pathnames, dummy URLs, user input, and Twitter handles. Here is an example: "Cypress performs most of its API tests via the `cy.request()` method, which serves as a `GET` command to the web server being tested."

A block of code is set as follows:

```
enum RedHatBoyState {
    Jumping,
    Running,
    Sliding,
}
```

When we wish to draw your attention to a particular part of a code block, the relevant lines or items are set in bold:

```
impl RedHatBoyContext {
        pub fn update(mut self, frame_count: u8) ->
        Self {
        ...
        self.position.x += self.velocity.x;
        self.position.y += self.velocity.y;

        if self.position.y > FLOOR {
            self.position.y = FLOOR;
        }
```

Any command-line input or output is written as follows:

```
the trait `From<SlidingEndState>` is not implemented for
`RedHatBoyStateMachine`
```

Bold: This indicates a new term, an important word, or words that you see onscreen – for instance, words in menus or dialog boxes appear in **bold**. Here is an example: "Upon any test launch from the GUI, users will have the ability to click on the **Add New Test** button."

> **Tips or Important Notes**
> Appear like this.

Get in touch

Feedback from our readers is always welcome.

General feedback: If you have questions about any aspect of this book, email us at customercare@packtpub.com and mention the book title in the subject of your message.

Errata: Although we have taken every care to ensure the accuracy of our content, mistakes do happen. If you have found a mistake in this book, we would be grateful if you would report this to us. Please visit www.packtpub.com/support/errata and fill in the form.

Piracy: If you come across any illegal copies of our works in any form on the internet, we would be grateful if you would provide us with the location address or website name. Please contact us at copyright@packt.com with a link to the material.

If you are interested in becoming an author: If there is a topic that you have expertise in and you are interested in either writing or contributing to a book, please visit authors.packtpub.com.

Share Your Thoughts

Once you've read, we'd love to hear your thoughts! Scan the QR code below to go straight to the Amazon review page for this book and share your feedback.

https://packt.link/r/1801070970

Your review is important to us and the tech community and will help us make sure we're delivering excellent quality content.

Part 1: Getting Started with Rust, WebAssembly, and Game Development

In this part, you'll build the skeleton for the application you'll be using for the rest of this book. You'll create your first WebAssembly app using Rust, and interact with JavaScript using `wasm-bindgen`. You'll also get started drawing to the Canvas, first with crude shapes and then with Sprites (image files) and even sprite sheets.

In this part, we cover the following chapters:

- *Chapter 1, Hello WebAssembly*
- *Chapter 2, Drawing Sprites*

1
Hello WebAssembly

Let's cut to the chase – if you're holding this book, you probably already know you love Rust, and you think **WebAssembly** is a great way to deploy your Rust programs to the web. Good news – you're right! Rust and WebAssembly are a match made in programmer heaven, and while WebAssembly is still in its early days, game development is an ideal candidate for WebAssembly. I am excited to be guiding you through building a game for the web in Stack Overflow's "most-loved" language, Rust.

This chapter is all about equipping yourself with the tools for the game development journey. In this chapter, we'll cover the following topics:

- What is WebAssembly?
- Creating a Rust and WebAssembly project skeleton
- Translating JavaScript code into Rust code
- Drawing to the screen with HTML5 Canvas

Technical requirements

To follow along with the project skeleton, you'll need to install `rustup` to install the Rust toolchains. This can be found at `https://rustup.rs/`. While you can install Rust and its various toolchains without using the `rustup` tool, it's not trivial, and I won't be documenting it here. You'll also need an editor for writing Rust code, and while you can use virtually any editor with rust-analyzer, if you're new to writing Rust, I'd recommend Visual Studio Code and the Rust extension found at `https://bit.ly/3tAUyH2`. It's easy to set up and works right out of the box.

Finally, you'll need a web browser, and in this chapter, you'll need some familiarity with the terminal and **Node.js**. If you get stumped, the code for this chapter is available at `https://github.com/PacktPublishing/Game-Development-with-Rust-and-WebAssembly/tree/chapter_1`. The final code for the entire book is in the main branch at `https://github.com/PacktPublishing/Game-Development-with-Rust-and-WebAssembly`.

Check out the following video to see the Code in Action: `https://bit.ly/3qMV44E`

What is WebAssembly?

You picked up this book (thanks!) so in all likelihood, you have some idea of what WebAssembly is, but just in case, let's grab a definition from `https://WebAssembly.org`:

> *"WebAssembly (abbreviated Wasm) is a binary instruction format for a stack-based virtual machine. Wasm is designed as a portable compilation target for programming languages, enabling deployment on the web for client and server applications."*

In other words, **Wasm** is a binary format that we can compile other languages to so that we can run them in the browser. This is different than transpiling or source-to-source compiling, where languages such as TypeScript are converted into JavaScript for running in JavaScript environments. Those languages are still ultimately running JavaScript, whereas Wasm is bytecode. This makes it a smaller download and parsing and compiling steps are removed when running it, which can lead to significant performance improvements. But let's be honest – you're not using Rust and Wasm for the performance improvements, which aren't guaranteed anyway. You're using it because you like Rust.

And that's okay!

Rust has a great type system, excellent developer tooling, and a fantastic community. While WebAssembly was originally created with C and C++ in mind, Rust is a fantastic language for WebAssembly for all the reasons you love Rust and more. Now, for most of the web's existence, writing applications to run in a browser meant writing JavaScript, and over the years, JavaScript has evolved into a suitably modern language for that purpose. I'm not here to tell you that if you like JavaScript you should stop, but if you love Rust, you should absolutely start compiling to Wasm and running apps in the browser.

> **Important Note**
>
> This book is focused on making web-based games with Rust and Wasm, but you can absolutely run Wasm apps in server-side environments such as Node.js. If you're interested in that, you can check out the book *Learn WebAssembly* by Mike Rourke, which can be found at `https://bit.ly/2N89prp`, or the official `wasm-bindgen` guide at `https://bit.ly/39WC63G`.

> **Important Note**
>
> This book assumes some familiarity with Rust, although you do not need to be an expert. If at any time you're confused by a Rust concept, I highly encourage you to stop and check *"the book"*, *The Rust Programming Language*, available for free at `https://doc.rust-lang.org/book/`.

So, now that I've convinced you to do what you were already going to do anyway, let's go over some of the tools you'll need to write a game for the web in Rust:

- `rustup`: Most likely you're already using `rustup` if you're writing Rust code. If you're not, you should, as it's the standard way to install Rust. It allows for easy installations of toolchains, Rust compilers, and even launches the Rust documentation. You'll need it to install the **Wasm toolchain**, and you can install it from the previous link. The code in this book has been tested on Rust version 1.57.0.

- **Node.js**: I know – I promised you that we'd be writing in Rust! We will, but this is still a web application and you'll be using Node.js to run the application. I recommend installing the current long-term support version (**16.13.0** at the time of writing). Older versions of Node.js may not work with the package creation tools as expected. If you're using Ubuntu Linux, be especially cautious when using the Debian distribution, which installs a very old version at this time. When in doubt, use tools for managing multiple versions, such as the **Node Version Manager** (**nvm**) tool for Linux/Mac or the corresponding nvm-windows tool for Windows, to ensure that you're using the long-term release version. I use the asdf tool (`https://asdf-vm.com/`) for managing multiple versions myself, although I don't usually recommend it to people that haven't used a version management tool before.

- **webpack**: We'll use webpack to bundle our application for release and run a development server. Most of the time, you won't have to worry about it, but it's there.

> **Important Note**
>
> The current template uses webpack 4. Make sure to check that when looking up documentation.

- `wasm-pack`: This is a Rust tool for building Rust-generated WebAssembly code. Like webpack, most of the time you won't know it's there, as it's managed by webpack, and your Rust application will largely be managed by Rust build tools.

- `wasm-bindgen`: This is one of the crates you'll need to get to know to write Rust-generated WebAssembly code. One limitation of WebAssembly is that you cannot access the **Document Object Model** (**DOM**) that represents a web page directly. Instead, WebAssembly programs need to call JavaScript functions to do that, requiring bindings and serializing data back and forth. What `wasm-bindgen` does is create those bindings and the boilerplate needed to call JavaScript functions from your Rust code, as well as provide tools to create bindings in the other direction so that JavaScript code can call back into the Rust code. We'll cover the details of how `wasm-bindgen` works as we go through the book, but to avoid getting bogged down in details right now, you can just think of it as a library to call JavaScript from your Rust code.

- `web-sys`: This is a crate made up of many pre-generated bindings, using `wasm-bindgen`, for the web. We'll use `web-sys` to call browser APIs such as the canvas and `requestAnimationFrame`. This book assumes at least a passing familiarity with web development but doesn't require expertise in this area, and in fact, one of the advantages of game development in Rust is that we can just treat the browser as a platform library that we call functions on. The `web-sys` crate means we don't have to create all those bindings ourselves.

- `Canvas`: HTML Canvas is a `<canvas>` browser element, such as headers or paragraphs, only it allows you to draw directly to it. This is how we can make a video game! There are many ways to draw to the canvas, including `WebGL` and `WebGPU`, but we're going to use the built-in Canvas API for most of this project. While this isn't the absolute fastest way of making a game, it's fast enough for learning purposes and avoids adding more technologies to our stack.

Finally, while googling `web-sys`, `web-bindgen`, or other Rust packages for WebAssembly, you are likely to come across references to `cargo-web` and `stdweb`. While both of those projects were important to the development of Rust as a WebAssembly source, neither has been updated since 2019 and can be safely ignored. Now that we know the tools we'll be using, let's start building our first Rust project.

A Rust project skeleton

> **Important Note**
>
> These directions are based on the status of `rust-webpack-template` at the time of writing. It's likely to have changed at the time of reading this, so pay close attention to the changes we are making. If they don't make sense, check the documents for `wasm-pack` and use your best judgment.

At this point, I'm going to assume you've installed `rustup` and Node.js. If you haven't, go ahead and follow the instructions for your platform to install them, and then follow these steps:

1. **Initialize the project**

    Let's start by creating a project skeleton for your application, which will be the Rust webpack Template from the Rust Wasm group. It's found on GitHub at `https://github.com/rustwasm/rust-webpack-template`, but you don't want to download it. Instead, use `npm init` to create it, like this:

    ```
    mkdir walk-the-dog
    cd walk-the-dog
    npm init rust-webpack
    ```

    You should see something like this:

    ```
    npx: installed 17 in 1.941s
       🦀 Rust + 🕸 WebAssembly + Webpack = ♡
    Installed dependencies ✔
    ```

    Congratulations! You have created your project.

2. **Install dependencies**

    You can install the dependencies with npm:

    ```
    npm install
    ```

    > **Important Note**
    >
    > If you prefer to use `yarn`, you can, with the exception of the `npm init` command. I'll be using `npm` for this book.

3. **Run the server**

After the installation completes, you can now run a development server with `npm run start`. You may see an error, like this:

```
i  Installing wasm-pack

Error: Rust compilation.
at ChildProcess.<anonymous> (/walk-the-dog/node_modules/@
wasm-tool/wasm-pack-plugin/plugin.js:221:16)
at ChildProcess.emit (events.js:315:20)
at maybeClose (internal/child_process.js:1048:16)
at Socket.<anonymous> (internal/child_process.js:439:11)
at Socket.emit (events.js:315:20)
at Pipe.<anonymous> (net.js:673:12)
```

If that happens, you'll need to install `wasm-pack` manually.

4. **Install wasm-pack**

On Linux and macOS systems `wasm-pack` is installed with a simple cURL script:

```
curl https://rustwasm.github.io/wasm-pack/installer/init.
sh -sSf | sh
```

Windows users have a separate installer that can be found at `https://rustwasm.github.io`.

5. **Run the server – take two**

Now that `wasm-pack` is installed, webpack can use it, and you should be able to run the app:

```
npm run start
```

When you see ⌈wdm⌋: `Compiled successfully.`, you can browse your app at `http://localhost:8080`. Okay, yes, it's a blank page, but if you open the developer tools console, you should see the following:

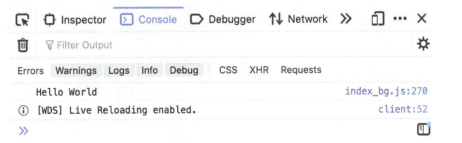

Figure 1.1 – Hello WebAssembly!

You've got the application running in the browser, but the Rust ecosystem updates faster than the template you used can keep up.

6. **Update the Rust edition**

The latest Rust edition, with the most recent Rust idioms and conventions, is 2021. This is changed in the generated `Cargo.toml` file in the `package` section, as shown here:

```
# You must change these to your own details.
[package]
name = "rust-webpack-template"
description = "Walk the Dog - the game for the Rust Games
with WebAssembly book"
version = "0.1.0"
authors = ["Eric Smith <paytonrules@gmail.com>"]
categories = ["wasm"]
readme = "README.md"
edition = "2021"
```

It is only the `edition` field that is changed here.

7. **Update the dependencies**

The dependencies in the generated `Cargo.toml` file are not going to be the latest and greatest unless you happened to pull the template down the moment it was updated. Since neither of us is that lucky, you're going to want to open up that file and modify the dependencies to the following. Please note that the ellipses are just there to mark a gap in the file and are not meant to be typed in:

```
wasm-bindgen = "0.2.78"
. . .
[dependencies.web-sys]
version = "0.3.55"
. . .
[dev-dependencies]
wasm-bindgen-test = "0.3.28"
futures = "0.3.18"
js-sys = "0.3.55"
wasm-bindgen-futures = "0.4.28"
```

Those are the versions I used while writing this book. If you're feeling adventurous, you can go to http://crates.io and find the most recent version of each dependency, which is what I would do, but I am a glutton for punishment. You're probably smarter than me and will use the versions specified here so that the sample code works.

8. **Update console_error_panic_hook**

   console_error_panic_hook is a very useful crate during the development of a WebAssembly application. It takes panics in Rust code and forwards them to the console so that you can debug them. The current template attempts to hide it behind a feature flag, but unfortunately, there's a bug and it doesn't work. Remember to double-check your generated code; if it doesn't look like what I've reproduced here, the bug may have been fixed, but in the meantime, delete the following code (still in Cargo.toml).

   ```
   [target."cfg(debug_assertions)".dependencies]
   console_error_panic_hook = "0.1.5"
   ```

   Then add the to the [dependencies] section, under wasm-bindgen is a good spot:

   ```
   console_error_panic_hook = "0.1.7"
   ```

Later, we'll make this a conditional dependency so that you don't deploy it during release builds, but for now, this is enough progress. Who wants to continue messing with config files anyway? I want to draw stuff to the screen!

> **Tip**
> While this application uses an npm init template to create itself, you can use its output to create a cargo generate template so that you don't have to redo these changes every time you create an application, simply by creating a git repository. Of course, if you do that, you'll fall behind changes to the rust-webpack template, so it's a trade-off. If you're curious about using cargo generate to create your own templates, you can find more information here: https://bit.ly/3hCFWTs.

Drawing to the canvas

To write our game in Rust, we're going to need to draw to the screen, and for that, we'll use the HTML Canvas element using the 2D context. What the canvas provides is an API for drawing directly to the screen, without knowledge of WebGL or using an external tool. It's not the fastest technology in the world but it's perfectly suitable for our small game. Let's start converting our Rust app from *"Hello World"* to an application that draws a **Sierpiński triangle**.

> **Important Note**
>
> The Sierpiński triangle is a fractal image that is created by drawing a triangle, then subdividing that triangle into four triangles, and then subdividing those triangles into four triangles, and so on. It sounds complicated but, as with many fractals, is created from only a few lines of math:

1. **Add the canvas**

   Canvas is an HTML element that lets us draw to it freely, making it an ideal candidate for games. Indeed, at the time of writing, Adobe Flash is officially dead, and if you see a game on the internet, be it 2D or 3D, it's running in a `canvas` element. Canvas can use WebGL or WebGPU for games, and WebAssembly will work quite well with those technologies, but they are out of the scope of this book. We'll be using the built-in Canvas 2D API and its 2D context. This means you won't have to learn a shading language, and we'll be able to get images on the screen very quickly. It also means that if you need to, you can find excellent documentation on the **Mozilla Developer Network** (**MDN**) Web Docs website: `https://mzl.la/3tX5qPC`.

   To draw to the canvas, we'll need to add it to the web page. Open up `static/index.html` and add underneath `<body>` tag `<canvas id="canvas" tabindex="0" height="600" width="600">Your browser does not support the canvas.</canvas>`. The width and height are pretty arbitrary but seem appropriate for now. The `"Your browser does not support the canvas."` message will show up on browsers that don't support HTML Canvas, but there aren't many of those anymore.

   > **Important Note**
   >
   > Make sure you don't delete the `<script>` tag. That's running the JavaScript and WebAssembly you're building in this project!

2. **Clean up errors**

Finally, we get to write some Rust code! Well, we get to delete some Rust code anyway. In the `src/lib.rs` file, you'll see a function named `main_js()` with the following code:

```
// This provides better error messages in debug mode.
// It's disabled in release mode so it doesn't bloat
   up the file size.
   #[cfg(debug_assertions)]
   console_error_panic_hook::set_once();
```

You can go ahead and remove the comments and the `[cfg(debug_annotations)]` annotation. For the time being, we'll leave that running in our build and will remove it when preparing for production with a feature flag.

> **Important Note**
>
> If you're seeing an error in your editor that says the `console::log_1(&JsValue::from_str("Hello world!"))` code is missing an unsafe block, don't worry – that error is wrong. Unfortunately, it's a bug in rust-analyzer that's been addressed in this issue: `https://bit.ly/3BbQ39m`. You'll see this error with anything that uses procedural macros under the hood. If you're using an editor that supports experimental settings, you may be able to fix the problem; check the `rust-analyzer.experimental.procAttrMacros` setting. When in doubt, check the output from `npm run start`, as that is the more accurate source for compiler errors.

> **Tip**
>
> If you diverge from this book and decide to deploy, go to *Chapter 10, Continuous Deployment*, and learn how to hide that feature behind a feature flag in release mode, so you don't deploy code you don't need into production.

Removing that code will remove the `warning: Found 'debug_assertions' in 'target.'cfg(...)'.dependencies'.` message on startup of the app. At this point, you may have noticed that I'm not telling you to restart the server after changes, and that's because `npm start` runs the `webpack-dev-server`, which automatically detects changes and then rebuilds and refreshes the app. Unless you're changing the webpack config, you shouldn't have to restart.

The current code

Up to now, I've been telling you what to do, and you've been blindly doing it because you're following along like a good reader. That's very diligent of you, if a little trusting, and it's time to take a look at the current source and see just what we have in our WebAssembly library. First, let's start with the `use` directives.

```
use wasm_bindgen::prelude::*;
use web_sys::console;
```

The first import is the `prelude` for `wasm_bindgen`. This brings in the macros you'll see shortly, and a couple of types that are pretty necessary for writing Rust for the web. Fortunately, it's not a lot, and shouldn't pollute the namespace too much.

> **Important Note**
>
> "**Pollute the namespace**" refers to what can happen when you use the `'*'` syntax and import everything from a given module. If the module has a lot of exported names, you have now those same names in your project, and they aren't obvious when you're coding. If, for instance, `wasm_bindgen::prelude` had a function named `add` in it and you also had a function named `add` in your namespace, they would collide. You can work around this by using explicit namespaces when you call the functions, but then why use * in the first place? By convention, many Rust packages have a module named `prelude`, which can be imported via * for ease of use; other modules should be imported with their full name.

The other import is `web_sys::console`, which brings in the `console` namespace from `web_sys`, which in turn mimics the `console` namespace in JavaScript. This is a good time to talk a little more in detail about what these two modules do. I've said it before but it probably bears repeating – `wasm_bindgen` provides the capability to bind JavaScript functions so you can call them in WebAssembly and to expose your WebAssembly functions to JavaScript. There's that language again, the one we're trying to avoid by writing Rust, but it can't be avoided because we're working in a browser.

In fact, one of the limitations of WebAssembly is that it cannot manipulate the DOM, which is a fancy way of saying that it can't change the web page. What it can do is call functions in JavaScript, which in turn do that work. In addition, JavaScript knows nothing about your WebAssembly types, so any data that is passed to a JavaScript object is marshaled into shared memory and then pulled back out by JavaScript so that it can turn it into something it understands. This is a LOT of code to write over and over again, and that is what the wasm-bindgen crate does for you. Later, we'll use it to bind our own custom bindings to third-party JavaScript code, but what about all the functions already built into the browser, such as console.log? That's where web-sys comes in. It uses wasm-bindgen to bind to all the functions in the browser environment so that you don't have to manually specify them. Think of it as a helper crate that says, *"Yeah, I know you'll need all these functions so I created them for you."*

So, to sum up, wasm-bindgen gives you the capability to communicate between WebAssembly and JavaScript, and web-sys contains a large number of pre-created bindings. If you're particularly interested in how the calls between WebAssembly and JavaScript work, check out this article by Lin Clark, which explains it in great detail, and with pictures: https://hacks.mozilla.org/2018/10/calls-between-javascript-and-webassembly-are-finally-fast-%F0%9F%8E%89/.

The wee allocator

After the use statements you'll see a comment block referring to the `wee_alloc` feature, which is a WebAssembly allocator that uses much less memory than the default Rust allocator. We're not using it, and it was disabled in the Cargo.toml file, so you can delete it from both the source code and Cargo.toml.

The main

Finally, we get to the main part of our program:

```
#[wasm_bindgen(start)]
pub fn main_js() -> Result<(), JsValue> {
```

The wasm_bindgen(start) annotation exports main_js so that it can be called by JavaScript, and the start parameter identifies it as the starting point of the program. If you're curious, you can take a look at pkg/index_bg.wasm.d.ts to see what was generated by it. You'll also want to take note of the return value, Result, where the error type can be JsValue, which represents an object owned by JavaScript and not Rust.

At this point, you may start to wonder how you'll keep track of what's JavaScript and what's Rust, and I'd advise you to not worry too much about it right now. There's a lot of jargon popping up and there's no way you'll keep it all in your head; just let it swim around in there and when it comes up again, I'll explain it again. JsValue is just a representative JavaScript object in your Rust code.

Finally, let's look at the contents:

```
console_error_panic_hook::set_once();

// Your code goes here!
console::log_1(&JsValue::from_str("Hello world!"));

Ok(())
```

The first line sets the panic hook, which just means that any panics will be redirected to the web browser's console. You'll need it for debugging, and it's best to keep it at the beginning of the program. Our one line, our *Hello World*, is console::log_1(&JsValue::from_ str("Hello world!"));. That calls the JavaScript console.log function, but it's using the version that's log_1 because the JavaScript version takes varying parameters. This is something that's going to come up again and again when using web-sys, which is that JavaScript supports varargs and Rust doesn't. So instead, many variations are created in the web-sys module to match the alternatives. If a JavaScript function you expect doesn't exist, then take a look at the Rust documents for web-sys (https://bit. ly/2N1RmOI) and see whether there are versions that are similar but built to account for multiple parameters.

> **Tip**
> A series of macros for several of the more commonly used functions (such as log) could solve this problem, but that's an exercise for the reader.

Finally, the function returns Ok(()), as is typical of Rust programs. Now that we've seen the generated code, let's break it down with our own.

Drawing a triangle

We've spent a lot of time digging into the code we currently have, and it's a lot to just write *"Hello World"* to the console. Why don't we have some fun and actually draw to the canvas?

What we're going to do is mimic the following JavaScript code in Rust:

```
canvas = window.document.getElementById("canvas")
context = canvas.getContext("2d")

context.moveTo(300, 0)
context.beginPath()
context.lineTo(0, 600)
context.lineTo(600, 600)
context.lineTo(300, 0)
context.closePath()
context.stroke()
context.fill()
```

This code grabs the canvas element we put in index.html, grabs its 2D context, and then draws a black triangle. One way to draw a shape on the context is to draw a line path, then stroke, and, in this case, fill it. You can actually see this in the browser using the web developer tools built into most browsers. This screenshot is from Firefox:

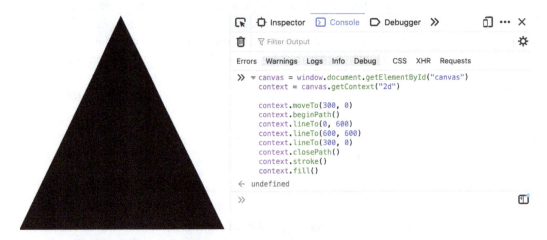

Figure 1.2 – A simple canvas triangle

Let's do the same thing in our Rust program. You'll see that it's a little…different. Start with the quick addition of a use statement at the top:

```
use wasm_bindgen::JsCast;
```

Then, replace the existing `main_js` function with the following:

```
console_error_panic_hook::set_once();

let window = web_sys::window().unwrap();
let document = window.document().unwrap();
let canvas = document
    .get_element_by_id("canvas")
    .unwrap()
    .dyn_into::<web_sys::HtmlCanvasElement>()
    .unwrap();

let context = canvas
    .get_context("2d")
    .unwrap()
    .unwrap()
    .dyn_into::<web_sys::CanvasRenderingContext2d>()
    .unwrap();

context.move_to(300.0, 0.0); // top of triangle
context.begin_path();
context.line_to(0.0, 600.0); // bottom left of triangle
context.line_to(600.0, 600.0); // bottom right of triangle
context.line_to(300.0, 0.0); // back to top of triangle
context.close_path();
context.stroke();
context.fill();
Ok(())
```

There are a few differences that stand out, but at a glance, you may just feel like Rust code is a lot *noisier* than JavaScript code, and that's true. You might be inclined to say that it's less elegant or isn't as clean, but I'd say that's in the eye of the beholder. JavaScript is a dynamically typed language and it shows. It ignores `undefined` and `null`, and can just crash if any of the values are not present. It uses duck typing to call all the functions on the context, which means that if the function is present, it simply calls it; otherwise, it throws exceptions.

Rust code takes a very different approach, one that favors explicitness and safety but at the cost of the code having extra noise. In Rust, you have to be more explicit when calling methods on structs, hence the casting, and you have to acknowledge `null` or failed `Result` types, hence all the unwraps. I've spent years using dynamic languages, including JavaScript, and I like them a lot. I certainly liked them a lot better than writing in C++, which I find overly verbose without really granting some of the safety advantages, but I think that with some tweaks, we can make Rust code nearly as elegant as JavaScript without glossing over exceptions and results.

My rant aside, if you're still running the program, you'll notice one minor detail – the Rust code doesn't compile! This leads me to the first thing we'll need to cover when translating JavaScript code to Rust code.

web-sys and feature flags

The `web-sys` crate makes heavy use of feature flags to keep its size down. This means that every time you want to use a function and it doesn't exist, you'll need to check which feature flag it's tied to, which is in its documentation, and add it to `Cargo.toml`. Fortunately, this is well documented and easy enough to do; we don't even need to restart the server!

Looking at our errors, we should see the following:

```
error[E0425]: cannot find function 'window' in crate 'web_sys'
--> src/lib.rs:18:27
  |
18 | let window = web_sys::window().unwrap();
  |                        ^^^^^^ not found in 'web_sys'
```

There are a few more errors of the same kind, but what you see here is that `window` is not in the `web_sys` module. Now, if you check the documentation for the `window` function in `web-sys` at `https://bit.ly/3ak3sAR`, you'll see that, yes, it does exist, but there is the `This API requires the following crate features to be activated: Window` message.

Open the `cargo.toml` file and look for `dependencies.web-sys`. You'll see that it has a `features` entry with just `["console"]` in it; go ahead and add `"Window"`, `"Document"`, `"HtmlCanvasElement"`, `"CanvasRenderingContext2d"`, and `"Element"` to that list. To be clear, you don't need all those feature flags for just the `window` function; that's all of the functions we're using.

You'll notice the project will rebuild automatically and should build successfully. If you look in the browser, you'll see your own black triangle! Let's extend it and learn a bit more about how we did it.

> **Tip**
> When a function you expect to exist on web-sys doesn't, go and check the feature flags in the documents.

DOM interaction

You'll notice that the method for drawing the triangle after you get the context looks essentially the same as the method in JavaScript – draw a line path, stroke, and fill it. The code at the top that interacted with the DOM looks…different. Let's break down what's going on here:

- Unwrapping option

  Getting the `Window` is just a function in the web-sys crate, one you enabled when you added the `Window` feature to `Cargo.toml`. However, you'll notice it's got unwrap at the end:

  ```
  let window = web_sys::window().unwrap();
  ```

  In JavaScript, `window` can be `null` or `undefined`, at least theoretically, and in Rust, this gets translated into `Option<Window>`. You can see that unwrap is applied to the result of `window()`, `document()`, and `get_element_by_id()` because all of them return `Option<T>`.

- `dyn_into`

  What the heck is `dyn_into`? Well, this oddity accounts for the difference between the way JavaScript does typing and the way Rust does. When we retrieve the canvas with `get_element_by_id`, it returns `Option<Element>`, and `Element` does not have any functions relating to the canvas. In JavaScript, you can use dynamic typing to assume the element has the `get_context` method, and if you're wrong, the program will throw an exception. This is anathema to Rust; indeed, this is a case where one developer's convenience is another developer's potential bug in hiding, so in order to use `Element`, we have to call the `dyn_into` function to cast it into `HtmlCanvasElement`. This method was brought into scope with the `use wasm_bindgen::JsCast` declaration.

> **Important Note**
> Note that `HtmlCanvasElement`, `Document`, and `Element` were all also feature flags you had to add in web-sys.

- Two unwraps?

  After calling `get_context("2d")`, we actually call `unwrap` twice; that's not a typo. What's going on is that `get_context` returns a `Result<Option<Object>>`, so we unwrap it twice. This is another case where the game can't recover if this fails, so `unwrap` is okay, but I wouldn't complain if you replaced those with `expect` so that you can give a clearer error message.

A Sierpiński triangle

Now let's have some real fun, and draw a Sierpiński triangle a few levels deep. If you're up for a challenge, you can try and write the code yourself before following along with the solution presented here. The way the algorithm works is to draw the first triangle (the one you are already drawing) and then draw another three triangles, where the first triangle has the same top point but its other two points are at the halfway point on each side of the original triangle. Then, draw a second triangle on the lower left, with its top at the halfway point of the left side, its lower-right point at the halfway point of the bottom of the original triangle, and its lower-left point at the lower-left point of the original triangle. Finally, you create a third triangle in the lower-right corner of the original triangle. This leaves a "hole" in the middle shaped like an upside-down triangle. This is much easier to visualize than it is to explain, so how about a picture?

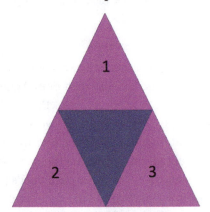

Figure 1.3 – A one-level Sierpiński triangle

Each of the numbered triangles was one that was drawn. The upside-down blue triangle is what's left of the original triangle because we didn't draw over it.

So that's one triangle subdivided into four. Now, the algorithm works recursively, taking each triangle and subdividing again. So, two levels deep, it looks like this:

Figure 1.4 – A two-level Sierpiński triangle

Note that it *doesn't* subdivide the upside-down triangle in the center, just the three purple ones that you created. Indeed, all the triangles with their points down are just "happy accidents" that make the shape look cool. You now know enough at this point to draw your own Sierpiński triangle, with one catch – you should remove the `fill` statement on context. Otherwise, all the triangles will be filled black and you won't be able to see them. Go ahead and give it a try.

Drawing the Sierpiński triangle

So, did you give it a try? No, I wouldn't either; I guess we have a lot in common. To get started with creating a Sierpiński triangle, let's replace the hardcoded triangle with a triangle function. Here's the first pass at `draw_triangle`:

```
fn draw_triangle(context: &web_sys::CanvasRenderingContext2d,
    points: [(f64, f64); 3]) {
        let [top, left, right] = points;
        context.move_to(top.0, top.1);
        context.begin_path();
        context.line_to(left.0, left.1);
        context.line_to(right.0, right.1);
        context.line_to(top.0, top.1);
        context.close_path();
        context.stroke();
}
```

There are a couple of small changes from the hard-coded version that we started with. The function takes a reference to the context and a list of three points. Points themselves are represented by tuples. We've also gotten rid of the `fill` function, so we only have an empty triangle. Replace the inline `draw_triangle` with the function call, which should look like this:

```
let context = canvas
    .get_context("2d")
    .unwrap()
    .unwrap()
    .dyn_into::<web_sys::CanvasRenderingContext2d>()
    .unwrap();

draw_triangle(&context, [(300.0, 0.0), (0.0, 600.0), (600.0,
600.0)]);
```

Now that you're drawing one empty triangle, you're ready to start drawing the recursive triangles. Rather than starting with recursion, let's draw the first subdivision by drawing three more triangles. The first will have the same top point and two side points:

```
draw_triangle(&context, [(300.0, 0.0), (150.00, 300.0), (450.0,
300.0)]);
```

Note that the third tuple has an x halfway between `300.0` and `600.0`, not between `0` and `600.0`, because the top point of the triangle is halfway between the other two points. Also note that y gets larger as you go down, which is upside-down compared to many 3D systems. Now, let's add the lower-left and lower-right triangles:

```
draw_triangle(&context, [(150.0, 300.0), (0.0, 600.0), (300.0,
600.0)]);
draw_triangle(&context, [(450.0, 300.0), (300.0, 600.0),
(600.0, 600.0)]);
```

Your triangles should look like this:

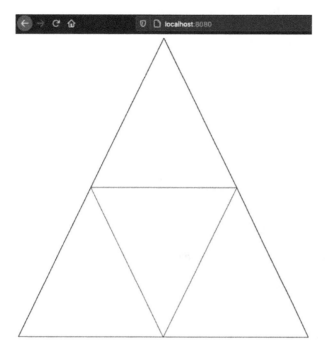

Figure 1.5 – Your triangles

You will start to see a pattern at this point, and we can begin to turn our hardcoded triangles into an algorithm. We'll create a function called `sierpinski` that takes the context, the triangle dimensions, and a depth function so that we only draw as many triangles as we want, instead of drawing them to infinity and crashing the browser. Then, we'll move those functions we called into that function:

```
fn sierpinski(context: &web_sys::CanvasRenderingContext2d,
points: [(f64, f64); 3], depth: u8) {
    draw_triangle(&context, [(300.0, 0.0), (0.0, 600.0),
    (600.0, 600.0)]);
    draw_triangle(&context, [(300.0, 0.0), (150.00, 300.0),
    (450.0, 300.0)]);
    draw_triangle(&context, [(150.0, 300.0), (0.0, 600.0),
    (300.0, 600.0)]);
    draw_triangle(&context, [(450.0, 300.0), (300.0,
    600.0), (600.0, 600.0)]);
}
```

This function currently ignores everything except the context, but you can replace those four `draw_triangle` calls from `main_js` and replace them with a call to `sierpinski`:

```
sierpinski(&context, [(300.0, 0.0), (0.0, 600.0), (600.0,
600.0)], 2);
```

It's important that you only send a depth of 2 for now so that the image will continue to look the same as we progress. Think of this call as a proto-unit test, guaranteeing our behavior doesn't change while we refactor. Now, in `sierpinski`, take the first triangle and have it use the passed-in points:

```
fn sierpinski(context: &web_sys::CanvasRenderingContext2d,
points: [(f64, f64); 3], depth: u8) {
    draw_triangle(&context, points);
    ...
```

Then, after drawing the triangle, reduce the depth by one and see if it is still greater than 0. Then, draw the rest of the triangles:

```
...
let depth = depth - 1;

if depth > 0 {
    draw_triangle(&context, [(300.0, 0.0), (150.00, 300.0),
        (450.0, 300.0)]);
    draw_triangle(&context, [(150.0, 300.0), (0.0, 600.0),
        (300.0, 600.0)]);
    draw_triangle(&context, [(450.0, 300.0), (300.0,
        600.0), (600.0, 600.0)]);
}
```

Now, to complete the recursion, you can replace all those `draw_triangle` calls with calls into `sierpinski`:

```
if depth > 0 {
    sierpinski(
        &context,
        [(300.0, 0.0), (150.00, 300.0), (450.0, 300.0)],
        depth,
```

```
    );
    sierpinski(
        &context,
        [(150.0, 300.0), (0.0, 600.0), (300.0, 600.0)],
        depth,
    );
    sierpinski(
        &context,
        [(450.0, 300.0), (300.0, 600.0), (600.0, 600.0)],
        depth,
    );
}
```

So far so good – you should still see a triangle subdivided into four triangles. Finally, we can actually calculate the midpoints of each line on the original triangle and use those to create the recursive triangles, instead of hardcoding them:

```
let [top, left, right] = points;
if depth > 0 {
    let left_middle = ((top.0 + left.0) / 2.0, (top.1 +
        left.1) / 2.0);
    let right_middle = ((top.0 + right.0) / 2.0, (top.1 +
        right.1) / 2.0);
    let bottom_middle = (top.0, right.1);
    sierpinski(&context, [top, left_middle, right_middle],
        depth);
    sierpinski(&context, [left_middle, left,
        bottom_middle], depth);
    sierpinski(&context, [right_middle, bottom_middle,
        right], depth);
}
```

Calculating the midpoint of a line segment is done by taking the x and y coordinates of each end, adding those together, and then dividing them by two. While the preceding code works, let's make it clearer by writing a new function, as shown here:

```
fn midpoint(point_1: (f64, f64), point_2: (f64, f64)) -> (f64,
f64) {
    ((point_1.0 + point_2.0) / 2.0, (point_1.1 + point_2.1)
    / 2.0)
}
```

Now, we can use that in the preceding function, for clarity:

```
if depth > 0 {
    let left_middle = midpoint(top, left);
    let right_middle = midpoint(top, right);
    let bottom_middle = midpoint(left, right);
    sierpinski(&context, [top, left_middle, right_middle],
      depth);
    sierpinski(&context, [left_middle, left,
      bottom_middle], depth);
    sierpinski(&context, [right_middle, bottom_middle,
      right], depth);
}
```

If you've been following along, you should make sure you're still showing a triangle with four inside to ensure you haven't made any mistakes. Now for the big reveal – go ahead and change the depth to 5 in the original Sierpinski call:

```
sierpinski(&context, [(300.0, 0.0), (0.0, 600.0), (600.0,
600.0)], 5);
```

You should see a recursive drawing of triangles, like this:

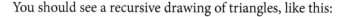

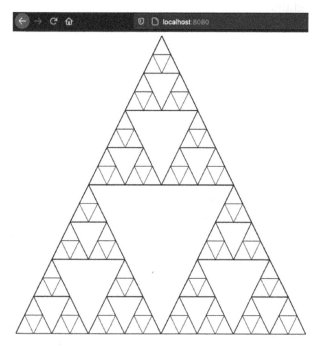

Figure 1.6 – A recursive drawing of triangles

Looking good! But what about those colors we saw in the original diagrams? They make it much more interesting.

When libraries aren't compatible

The earlier examples of this image had the triangles filled in with a different random color at each recursive layer. So, the first triangle was one color, three and four were another, the next nine another, and so on. It makes for a more interesting image *and* it provides a good example of what to do when a library isn't completely WebAssembly-compatible.

To create a random color, we'll need a random number generator, and that is not part of the standard library but instead found in a crate. You can add that crate by changing the Cargo.toml file to include it as a dependency:

```
console_error_panic_hook = "0.1.7"
rand = "0.8.4"
```

When you do this, you'll get a compiler error that looks like the following (although your message may differ slightly):

```
error: target is not supported, for more information see:
https://docs.rs/getrandom/#unsupported-targets
    --> /usr/local/cargo/registry/src/github.com-
    1ecc6299db9ec823/getrandom-0.2.2/src/lib.rs:213:9
    |
213 | /          compile_error!("target is not supported, for
more information see: \
214 | |                                  https://docs.rs/
getrandom/#unsupported-targets");
```

This is a case where a transitive dependency, in this case getrandom, does not compile on the WebAssembly target. In this case, it's an extremely helpful error message, and if you follow the link, you'll get the solution in the documentation. Specifically, you need to enable js in the feature flags for getrandom. Go back to your Cargo.toml file and add the following:

```
getrandom = { version = "0.2.3", features = ["js"] }
```

This adds the getrandom dependency with the js feature enabled, and your code will begin compiling again. The lesson to take away from this is that not every Rust crate will compile on the WebAssembly target, and when that happens, you'll need to check the documents.

> **Tip**
> When a crate won't compile *slowly*, read the error message and follow the instructions. It's very easy to skim right over the reason the build is breaking, especially when you're frustrated.

Random colors

Now that we've got the random create building with our project, let's change the color of the triangles as we draw them to a random color. To do that, we'll set fillStyle with a color before we draw the triangle, and we'll add a fill command. This is, generally, how the Context2D API works. You set up the state of the context and then execute commands with that state set. It takes a little getting used to but you'll get the hang of it. Let's add color as a parameter of the three u8 tuples to draw_triangle:

```
fn draw_triangle(
    context: &web_sys::CanvasRenderingContext2d,
```

```
    points: [(f64, f64); 3],
    color: (u8, u8, u8),
) {
```

> **Important Note**
>
> Colors are represented here as three components, red, green, and blue, where each value can go from 0 to 255. We're using tuples in this chapter because we can make progress quickly, but if it's starting to bother you, you're welcome to make proper `structs`.

Now that `draw_triangle` needs a color, our application doesn't compile. Let's move to the `sierpinski` function and pass a color to it as well. We're going to send the color to the `sierpinski` function, instead of generating it there, so that we can get one color at every level. The first generation will be one solid color, then the second will all be one color, and then the third a third color, and so on. So let's add that:

```
fn sierpinski(
    context: &web_sys::CanvasRenderingContext2d,
    points: [(f64, f64); 3],
    color: (u8, u8, u8),
    depth: u8,
) {
    draw_triangle(&context, points, color);
    let depth = depth - 1;

    let [top, left, right] = points;
    if depth > 0 {
        let left_middle = midpoint(top, left);
        let right_middle = midpoint(top, right);
        let bottom_middle = midpoint(left, right);
        sierpinski(&context, [top, left_middle,
          right_middle], color, depth);
        sierpinski(&context, [left_middle, left,
          bottom_middle], color, depth);
        sierpinski(&context, [right_middle, bottom_middle,
          right], color, depth);
    }
}
```

I put `color` as the third parameter and not the fourth because I think it looks better that way. Remember to pass color to the other calls. Finally, so that we can compile, we'll send a color to the initial `sierpinski` call:

```
sierpinski(
    &context,
    [(300.0, 0.0), (0.0, 600.0), (600.0, 600.0)],
    (0, 255, 0),
    5,
);
```

Since this is an RGB color, (0, 255, 0) represents green. Now, we've made our code compile, but it doesn't do anything, so let's work back downward from the original call and into the `sierpinski` function again. Instead of just passing the color through, let's create a new tuple that has a random number for each component. You'll need to add `use rand::prelude::*;` to the use declarations at the top. Then, add the following code to the `sierpinski` function, after the `if depth > 0` check:

```
let mut rng = thread_rng();

let next_color = (
    rng.gen_range(0..255),
    rng.gen_range(0..255),
    rng.gen_range(0..255),
);

...
sierpinski(
    &context,
    top, left_middle, right_middle],
    next_color,
    depth,
);
sierpinski(
    &context,
    [left_middle, left, bottom_middle],
    next_color,
```

```
        depth,
    );
sierpinski(
        &context,
        [right_middle, bottom_middle, right],
        next_color,
        depth,
    );
```

Inside the depth check, we randomly generate `next_color` and then pass it along to all the recursive `sierpinski` calls. But of course, our output *still* doesn't look any different. We never changed `draw_triangle` to change the color! This is going to be a little weird because the `context.fillStyle` property takes `DOMString` in JavaScript, so we'll need to do a conversion. At the top of `draw_triangle`, add two lines:

```
let color_str = format!("rgb({}, {}, {})", color.0, color.1,
color.2);
context.set_fill_style(&wasm_bindgen::JsValue::from_str(&color_
str));
```

On line one, we convert our tuple of three unsigned integers to a string reading `"rgb(255, 0, 255)"`, which is what the `fillStyle` property expects. On the second line, we use `set_fill_style` to set it, doing that funky conversion. There are two things that you need to understand with this function. The first is that, generally, JavaScript properties are just public and you set them, but `web-sys` generates `getter` and `setter` functions. The second is that these generated functions frequently take `JsValue` objects, which represent an object owned by JavaScript. Fortunately, `wasm_bindgen` has factory functions for these, so we can create them easily and use the compiler as our guide.

> **Tip**
> Whenever you translate from JavaScript code to Rust, make sure that you check the documentation of the corresponding functions to see what types are needed. Passing a string to JavaScript isn't always as simple as you might think.

Finally, we actually need to fill the triangles to see those colors, so after `context.stroke()`, you need to restore that `context.fill()` method you deleted earlier, and ta-da!

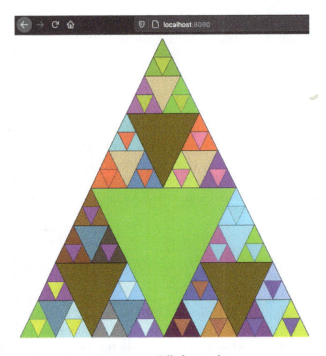

Figure 1.7 – Filled triangles

You've done it, and you're ready to start creating a real game.

Summary

In this chapter, we've done a *lot*. We've written our first WebAssembly app using Rust, moving from *"Hello World"* to drawing in the browser with HTML Canvas. You've added crates, run a development server, and interacted with the DOM. You've learned a lot about interacting with the browser, including the following:

- Creating the main entry point with `#[wasm_bindgen(start)]`
- Translating JavaScript code to Rust code
- Dealing with crates that compile to JavaScript

You've also been introduced to HTML Canvas. Frankly, it's been a bit of a whirlwind, so don't worry if some information flew over your head, as we'll cover many of these topics again – including in the next chapter, where we'll start drawing sprites.

2
Drawing Sprites

Now that we've got a working app and we're drawing to the screen, we can start making something that actually looks like a game. That means **rendering sprites**, which is just a fancy way of saying drawing pictures. So, in this chapter, we'll start by defining what those pictures are by doing a little bit of game design, and then we'll render a static sprite to the screen. Since a static picture is a pretty boring game, we'll even get the sprite animating too.

In this chapter, we'll do the following:

- Design our game, Walk the Dog.
- Render a sprite to the Canvas.
- Use a sprite sheet to load many sprites at once.
- Animate a character via the sprite sheet.

By the end of this chapter, you'll be drawing characters instead of static triangles, and you'll even have them running on the screen.

Technical requirements

In addition to the technical requirements of *Chapter 1, Hello WebAssembly*, you'll need to download the assets found at `https://github.com/PacktPublishing/Game-Development-with-Rust-and-WebAssembly/wiki/Assets`. We'll build on top of the results of that chapter as well, so don't throw away the code. If you're reading this book out of order because you can't be tamed by society's rules, then you can get the previous chapter's source code at `https://github.com/PacktPublishing/Game-Development-with-Rust-and-WebAssembly/tree/chapter_1` and start there. If you get stumped, you can find the complete source code for this chapter at `https://github.com/PacktPublishing/Game-Development-with-Rust-and-WebAssembly/tree/chapter_2`.

Check out the following video to see the Code in Action: `https://bit.ly/3wOpCqy`

A quick game design session

In the previous chapter, I had you create a project called "Walk the Dog", and you were so engrossed by the process of creating a Rust project and my thrilling prose that you didn't even ask why that was the name of the project. Now we'll dig into the game we're making for this book – **Walk the Dog**.

Walk the Dog is an endless runner with a simple concept. You play as a boy walking his dog through the forest when your dog is surprised by a cat that runs by and starts chasing it. You, in turn, begin chasing your dog through the forest, dodging obstacles along the way, until you crash into one and fall down. At which point, of course, the dog turns around and checks on you.

In case you hadn't guessed, the idea for this game came to me while walking the dog on ice. I've used *Miro* (`https://miro.com`) to make a prototype, just to get a feel for what the game will look like:

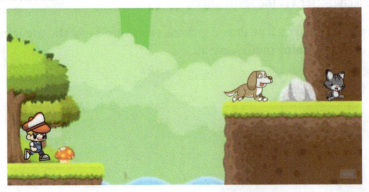

Figure 2.1 – A Walk the Dog screen, hypothetically

Before you get the idea that I'm a great artist, all of the assets I'm using are freely available online via creative commons licenses. You might notice that the background is a little fuzzy relative to the characters, and that's because I made almost no effort to scale the characters to fit beyond copying and pasting them into Miro and dragging the corners around. When we place the actual objects in our game, we'll need to make a better effort than that.

The temptation at this point is to say, "we're done" and start coding. Given the small size of our game, I don't think we need a full treatment to start coding, but I do want to make sure that we clarify a few things about the game.

Scoring is done by measuring how far our little **Red Hat Boy** (**RHB** for short) runs – the same as most endless runners such as *Canabalt* (`http://canabalt.com/`) or the *Dinosaur T-Rex* game that shows up when you start Google Chrome without an internet connection. The dog and cat navigate all obstacles effortlessly and are just there to give the player ideas on how to catch the dog, and perhaps mislead the player by taking a path they cannot follow. Obstacles will include rocks and boxes that you can crash into and water that you can fall into. RHB has a slide animation, so sometimes he'll need to slide under little cliffs too, which the dog runs under effortlessly. It's not enough for a fully fledged game, but it's enough to give us a checklist of features for future chapters. Let's say goodbye to our lovely triangles and begin rendering our adorable Red Hat Boy.

Rendering a sprite

Sprite is a term so commonplace that it's possible to use it in conversation without actually knowing its meaning, yet properly defining it means properly defining bitmap, which in turn means properly defining pixmap. Did you know the term sprite was coined in the 1970s by Danny Hillis (`http://bit.ly/3aZlJ72`)? It's exhausting.

While I find all of this fascinating, you didn't get this book for that, so for our purposes, a sprite is a 2D image loaded from a file. Red Hat Boy, his dog and cat, and the background will all be sprites. Let's not waste any more time on definitions and start drawing one.

Loading images

We'll start by unzipping the assets and copying the `Idle (1).png` file from `resized/rhb` into the `static` directory in your project. This will make it reachable from your program. As we build the program out, we'll need further organization, but for one file, this is fine. Next, we'll need to modify our code. You can leave the Sierpiński triangle in there for now as it looks cute next to the sprite, but the first thing to do is use the `HTMLImage` element to load an image. For now, it's important that you load and draw the image **before** calling into the Sierpiński triangle. It looks like this:

```
#[wasm_bindgen(start)]
pub fn main_js() -> Result<(), JsValue> {
    ....
    let image = web_sys::HtmlImageElement::new().unwrap();
    image.set_src("Idle (1).png");
    sierpinski(
        &context,
        [(300.0, 0.0), (0.0, 600.0), (600.0, 600.0)],
        (0, 255, 0),
        5,
    );
    Ok(())
}
```

You will once again get the `^^^^^^^^^^^^^^^^^ could not find `HtmlImageElement` in `web_sys`` error. Remember that the web-sys crate makes heavy use of feature flags, so you'll need to add `HtmlImageElement` to the feature flag list in `Cargo.toml`. After you add that, rebuilding will take a little longer, but the application will build again. Now you have loaded the image, and we can draw it.

Canvas coordinates

Before we draw it, we need to cover one thing about the **Canvas** that you might have noticed from the first chapter, and that is the coordinate system. In the first chapter, we covered how the Canvas has a context; in fact, it has multiple contexts, but you can only use one per canvas, and that we are using the **2D** context, which gives us an API for drawing directly to the screen. Then, we made a bunch of `line_to` and `move_to` commands that may not have made sense at the time, which is why we need to discuss the coordinate system:

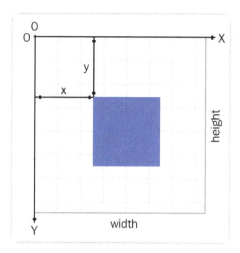

Figure 2.2 – Source: Mozilla (http://mzl.la/30NLhxX)

Our canvas is divided into a 2D grid with dimensions of 600 by 600. Why 600 x 600? Because that's the height and width of our canvas element on the HTML page that we created in *Chapter 1, Hello WebAssembly*. The size itself was completely arbitrary, and we'll probably change it as our game grows. The units of the grid are pixels, so when we moved the top of our original triangle to (300.0, 0.0), we moved it 300 pixels to the right (because **x** is first) and 0 pixels down (because **y** is second). Note that **0** is at the top of the screen because (0.0, 0.0) is in the top-left corner of the canvas.

Drawing images

Drawing one image at this point seems uncomplicated – we'll use the drawImage command from JavaScript; only we'll use the web-sys version for HtmlElement.

Tip

Remember that JavaScript functions frequently use function overloading, which Rust doesn't support, so one JavaScript function may have many corresponding variations in Rust.

So, let's add the draw command right after the code to load the image, and we'll be done:

```
image.set_src("Idle (1).png");
context.draw_image_with_html_image_element(&image, 0.0, 0.0);
...
```

We've ignored `Result` from the `draw_image_with_html_image_element` command, but that should draw the image, except, it…doesn't. It turns out you can't draw the image immediately after setting the source of an image element because the image hasn't been loaded yet. In order to wait for the image to be loaded, we'll use the `onload` callback of `HtmlImageElement`, which you can set up using `set_onload` in Rust. To do that, you'll need to learn a little about using JavaScript callbacks from Rust in the WebAssembly environment.

JavaScript callbacks

When you set the `onload` callback via the `set_onload` function in Rust, you're calling into JavaScript from WebAssembly, via a function that `web-sys` has generated for you. Unfortunately, translating the following JavaScript to Rust is complicated by the fact that JavaScript is garbage collected whereas Rust uses manual memory management, with its famous borrow checker. Take this code for example:

```
image.onload = () => { alert("loaded"); }
```

What this means is to actually pass a function to JavaScript, as we want to do here, you have to use a complicated signature as well as think carefully about the borrowing rules of Rust. It's the kind of code that finally makes sense after you get it right but can be hard to write. Let's work through what we need to do here.

Back in our source code, right after creating `HtmlImageElement`, we can try to add an `onload` callback in a way that seems intuitive:

```
let image = web_sys::HtmlImageElement::new().unwrap();
image.set_onload(|| {
    web_sys::console::log_1(&JsValue::from_str("loaded"));
});
image.set_src("Idle (1).png");
...
```

Intuitive might be an exaggeration, but that fits with the code we know how to write so far. Unfortunately, that doesn't work, as you'll get compiler errors about mismatched types, as shown here:

```
error[E0308]: mismatched types
  --> src/lib.rs:43:22
   |
   |            image.set_onload(|| {
```

```
|  _____^
|  |            web_sys::console.log_1("loaded");
|  |        });
|  |_____^ expected enum `Option`, found closure
|
= note: expected enum `Option<&js_sys::Function>`
        found closure `[closure@src/lib.rs:43:22: 45:6]`
```

As the error says, set_onload doesn't take a Rust closure but instead takes
Option<&js_sys::Function>. Unfortunately, the error doesn't tell you how to fix it,
and it's not clear how to create the js_sys::Function object. What you can do is start
by creating a Closure object, with a capital "C," and try passing that into set_onload:

```
let image = web_sys::HtmlImageElement::new().unwrap();
```

```
let callback = Closure::once(|| {
    web_sys::console::log_1(&JsValue::from_str("loaded"));
});
image.set_onload(callback);
```

Closure is a wasm-bindgen struct that is used to transfer a Rust closure to JavaScript.
Here, we are using the once function on Closure because we know the onload
handler is only called once. However, we still can't just send that to JavaScript as is; trying
to via image.set_onload(callback) results in the following error:

```
error[E0308]: mismatched types
  --> src/lib.rs:47:22
   |
47 |        image.set_onload(callback);
   |                         ^^^^^^^^ expected enum `Option`,
   found struct `wasm_bindgen::prelude::Closure`
```

Remember that set_onload wants Option<&js_sys::Function>, and so far,
we've only created Closure. Fortunately, the Closure struct provides a way to make
that conversion, which looks like this:

```
image.set_onload(Some(callback.as_ref().unchecked_ref()));
```

First, we call `as_ref` on the callback, which returns a raw `JsValue`, and then we call `unchecked_ref`, which converts it to a `&Function` object. We pass that into `Some` because `onload` can be `null` in JavaScript. Hooray! It compiles! The draw code now looks like this:

```
let image = web_sys::HtmlImageElement::new().unwrap();

let callback = Closure::once(|| {
    web_sys::console::log_1(&JsValue::from_str("loaded"));
});

image.set_onload(Some(callback.as_ref().unchecked_ref()));

image.set_src("Idle (1).png");
context.draw_image_with_html_image_element(&image, 0.0, 0.0);
...
```

And if you run the app, it **still** doesn't show our image, but it does log an error in the browser's console:

```
Uncaught Error: closure invoked recursively or destroyed
already
```

When was the closure destroyed? This is all in the `main` function, and so the closure is destroyed a couple of lines later when the function completes and the `callback` variable is no longer in scope. In order to see our log message, we can add one more call to the code after setting the `on_load` function:

```
image.set_onload(Some(callback.as_ref().unchecked_ref()));
callback.forget();
```

Calling `forget()` on the callback hands over memory management from Rust to JavaScript, effectively creating a deliberate memory leak. This is not something we want to do often, and it's here strictly to get us past our latest error, by preventing the closure from being destroyed. If you compile that and check your browser console, you'll see the message "loaded" now. This is great and all, but it still doesn't draw our picture because we're not actually waiting for the image to be loaded yet. For that, we'll need an asynchronous function.

> **Tip**
> Translating Rust closures to JavaScript closures is one of those cases where the abstractions between JavaScript and Rust are leaking all over the place, and it can be "accidentally-swear-in-front-of-your-kids" frustrating. So, don't feel bad when you get this wrong or get confused; it just means you're human.

> **Important Note**
> We'll have more examples of sending closures to JavaScript throughout this book, but you may find yourself wanting to cross-reference the official documents at `http://bit.ly/3kSyOSI` and `http://bit.ly/3sXt1OW`.

Async Rust

Rust added **async/.await** support in version 1.39 as a "syntactic sugar" around futures. If you haven't used them before (in Rust or other languages), the shorthand is that you can annotate a function or block with `async`, letting the runtime know that a function can be made to run asynchronously. Inside that function, you can then use an `await` call to pause execution of the current thread/process and allow the rest of the application to continue until the function that's being waited for can resume. The key thing is that while `await` pauses execution of the current execution context, it allows the rest of the code to continue. This makes it ideal for a game that cannot ever stop executing its game loop. It's also significantly cleaner to work with than callback-based code, so we'll be using it here (in combination with channels) to make sure that we don't try to draw our image before it's loaded.

If you're familiar with using `async/.await` in traditional Rust, then you know these functions need to execute in a runtime, usually using crates such as `tokio` or `async-std`. That runtime is responsible for handing off control and restoring it. Take this simple example from the `async-std` library's repository:

```
use async_std::task;

fn main() -> Result<(), surf::Error> {
    task::block_on(async {
        let url = "https://www.rust-lang.org";
        let mut response = surf::get(url).send().await?;
        let body = response.body_string().await?;
```

```
        dbg!(url);
        dbg!(response.status());
        dbg!(response.version());
        dbg!(response.header_names());
        dbg!(response.header_values());
        dbg!(body.len());

        Ok(())
    })
}
```

Here, the `async` block is wrapped in a function called `task::block_on`, which handles stopping the execution of this block on each `await` call and then resuming execution when the `await` "wakes up" for future processing. All of this requires spawning threads or checking event loops, code that you don't have to write because you're getting it from `async-std`.

If you're familiar with other languages that natively support async/.await syntax, such as JavaScript, you might wonder why this additional step is necessary. The answer is that, unlike JavaScript, Rust doesn't **have** a runtime, asynchronous or otherwise, so we have to provide one. It's a little uncomfortable that the `async` and `await` keywords exist in the language but don't work without additional crates, but that's a price we pay for additional power.

That's the bad news, but now for the good news – in WebAssembly, we don't need any additional runtime! Our code runs in the browser and can, therefore, use the browser's runtime; we just need to use a crate to spawn futures on the local event loop, and that crate is already present – `wasm_bindgen_futures`.

Spawning a future

A **future**, in Rust and other languages, is a data structure that represents an asynchronous computation. In other words, it's a reference to something that will happen *in the future*, hence the name. In the case of Rust, it's actually implemented as a trait. In Rust, the trait is named `Future`, naturally, but as a user of futures, you don't generally create the `Future` type directly. You declare a function or closure as `async`, and when an `async` function is called, its return value will be wrapped in `Future`. Then, the caller can wait for that `Future` instance to finish by calling `await`. The benefit of this approach is that while the program doesn't actually stop when you call `await`, it looks like it does from the perspective of the code author. This makes the code look a lot more linear. In reality, execution of your program continues; otherwise, it would become unresponsive, but the runtime handles picking up the program where it left off when `Future` completes.

In case you forgot, we're trying to draw a sprite to the canvas, and to do so, we have to wait for the image to be loaded first. For that, we'll eventually use futures, but there's some infrastructure we need to build first. We'll start by adding an `onload` callback to `HtmlImageElement`, which will call a `oneshot` channel when the image is loaded. A `oneshot` channel is a channel where the receiver implements the `Future` trait, so we can call `await` on it to wait for it to receive a message. If we set up the `onload` callback to send a message to that channel, we can then call `await` on the receiver so execution will block until that image is loaded. Then, we can actually draw the image, knowing it's loaded. For all that to work, we need to wrap everything in an `async` block and spawn the returned future. That's a limitation of the `await` syntax; it can only work inside an `async` function of a block. Naturally, we'll start the implementation in the...`Cargo.toml` file.

I guess it's not natural to start with the .TOML file, but we need to pull the future dependencies into our WebAssembly project. They're already present in testing, so we'll move `futures` and `wasm-bindgen-futures` out of `dev-dependencies` and into the standard `dependencies` block. You can put them right under `getrandom`, as shown in the following code:

```
getrandom = { version = "0.2.3", features = ["js"] }
futures = "0.3.17"
wasm-bindgen-futures = "0.4.28"
```

Now that we have access to Rust futures, we can use `wasm_bindgen_futures::spawn_local` to spawn a local future and put all our code for drawing the image into it. Returning to the code we wrote earlier to load `HtmlImageElement`, we will want to wrap all of it in a call to `spawn_local`, as shown in the following code:

```
wasm_bindgen_futures::spawn_local(async move {
    let image = web_sys::HtmlImageElement::new().unwrap();

    let callback = Closure::once(move || {
        web_sys::console::log_1(&JsValue::from_str("loaded"));
    });

    image.set_onload(Some(callback.as_ref().unchecked_ref()));
    callback.forget();

    image.set_src("Idle (1).png");
    context.draw_image_with_html_image_element
        (&image, 0.0, 0.0);
    sierpinski(
    ...
});
```

When you call `spawn_local`, you need to pass it as a block marked with `async`, because `spawn_local` requires `Future`. We've marked this block as move to give the block ownership of any bindings we reference in it. Later, we'll also need to make sure we deal properly with the lifetime of this closure, which must be `'static`, but right now, we don't have to worry about that because everything is in the closure. This image is still not going to draw because while `Future` gets spawned when it runs to completion, our program exits. We need to wait for the image to be loaded, and for that, we'll use the `oneshot` channel.

A `oneshot` channel works like its name; you can call once, at which point it is consumed and cannot be called again. This means that when you move a `oneshot` channel into a Rust closure, the closure immediately becomes `FnOnce`. In fact, if you try to move `oneshot` into `FnMut` or `Fn`, you'll get compiler errors, with a side effect of your hair falling out as you try to figure out what's wrong. So, don't do that – it hurts.

Instead, let's create the channel inside the `spawn_local` block, and then replace the `web_sys::console::log_1` call in the callback by sending a message to the channel. The changes are shown as follows:

```
let (success_tx, success_rx) =
futures::channel::oneshot::channel::<()>();

let image = web_sys::HtmlImageElement::new().unwrap();


let callback = Closure::once(move || {
    success_tx.send(());
});

...
```

On the first line, we created the `oneshot` channel of `unit` type and then moved its transmitter into the callback. We got rid of the log message and replaced it with a call to `send` on the transmitter. Now, we need to wait for that message to be sent before we try to draw the image. Let's modify the code underneath the closure:

```
image.set_onload(Some(callback.as_ref().unchecked_ref()));
image.set_src("Idle (1).png");


success_rx.await;
context.draw_image_with_html_image_element(&image, 0.0, 0.0);
```

First, we remove the `forget` call, as it's no longer necessary since we'll wait for the `onload` function to be called before we try to draw the image. That makes it okay for the closure to be deleted when the scope completes. Then, we call `success_rx.await` to block while the load completes. Finally, we'll draw the image as we were doing before, and it shows up!

Figure 2.3 – It is I, Red Hat Boy, king of the triangle

Important Note

We're ignoring a lot of results here, which is a bad practice. In the next chapter, we'll start structuring our game to better separate concerns, and when doing so, we'll remove that in favor of explicit error handling or calls to `expect` if we truly want to halt execution.

You might wonder why we're using `spawn_local` at all here, instead of just using a standard Rust channel and calling `recv` on it, and the reason is that a `recv` call blocks the main thread's execution, which is a big no-no in the browser. Browser-based code must allow the web browser to continue its event loop, and pausing it will cause the browser itself to become unresponsive. You could use a `try_rcv` call because that doesn't block, but you'd have to check it in a loop to make sure to wait until the image loads. That would also pause the browser and would probably cause one of those irritating `The browser is not responding` errors. Since both the browser and video games coincidentally can't cause the app to become unresponsive, we will use a `spawn_local` block and the `async/await` syntax. Remember that while the `await` context pauses local execution, the program itself actually keeps running, if only to constantly poll and see whether `Future` is complete.

Congratulations! You've drawn an image on the screen a mere thousand words after I promised you would, but there's one more thing we have to concern ourselves with. Let's make a small change to the code:

```
image.set_onload(Some(callback.as_ref().unchecked_ref()));
image.set_src("rhg.png");
success_rx.await;
```

Now, if you run the app, the screen doesn't draw anything, including the triangle! This is because we await a successful load, which will never come. We need to handle the error case as well so that we can continue in the event of a failed load rather than hanging, even if we just want to halt the error. What we want to do is send one message (a unit) on success and another message (the error) to the receiver when the image is finished loading, either one way or the other.

You might think you can change `success_tx` to take either `unit` or an error code when the load fails. We can use `JsValue` for the error since that's the type any error code from the browser will be.

> **Important Note**
>
> `JsValue` is a type that represents any value that comes directly from JavaScript. In Rust code, we'll frequently convert these types into more specific Rust types.

That code will look like this:

```
let (success_tx, success_rx) =
futures::channel::oneshot::channel::<Result<(), JsValue>>();
let image = web_sys::HtmlImageElement::new().unwrap();

let callback = Closure::once(move || {
    success_tx.send(Ok(()));
});
let error_callback = Closure::once(move |err| {
    success_tx.send(Err(err));
});


image.set_onload(Some(callback.as_ref().unchecked_ref()));
image.set_onerror(Some(error_callback.as_ref().unchecked_
ref()));
```

This is immediately going to be a compiler error:

```
70 |             let error_callback = Closure::once(move |err| {
   |
^^^^^^^^^^ value used here after move
71 |                 success_tx.send(Err(err));
```

success_tx cannot be moved into both closures at the same time. We're going to need to use one of the Rust constructs to share the channel across threads so that we can use it in both callbacks.

> **Important Note**
>
> We could use two oneshot channels and a select statement here as well, but at the time of writing, that did not work well in WebAssembly.

What we'll do is create the channel and then set up a reference counted version of the success and error transmitters. This means that both transmitters will send to the same receiver. Both of these will need to wrapped inside Mutex, as shown here, replacing the original creation of the oneshot channel:

```
let (success_tx, success_rx) =
futures::channel::oneshot::channel::<Result<(), JsValue>>();
```

```
let success_tx = Rc::new(Mutex::new(success_tx));
let error_tx = Rc::clone(&success_tx);
```

Note that we will start sending the channel the Result, so we can differentiate between success and failure later. You'll need to make sure that you import std::rc::Rc and std::sync::Mutex. Now that success_tx has been changed into Rc<Mutex<Sender>>, you'll need to update the success callback to reflect it. You'll want to lock access to Mutex and then send the success message. Your first try is likely to look like this:

```
let image = web_sys::HtmlImageElement::new().unwrap();
let callback = Closure::once(move || {
    success_tx
        .lock()
        .and_then(|oneshot| Ok(oneshot.send(Ok(()))));
});
...
```

This locks Mutex and then sends its oneshot an Ok(()). This is *almost* right, but there's a problem leading to a compiler error, as shown here:

```
error[E0507]: cannot move out of dereference of
`std::sync::MutexGuard<'_, futures::futures_
channel::oneshot::Sender<Result<(), wasm_bindgen::JsValue>>>`
  --> src/lib.rs:38:40
   |
38 |                        .and_then(|oneshot| Ok(oneshot.
send(Ok(())))));
   |
^^^^^^^^^^^^^^^^^^^^ move occurs because value has type
`futures::futures_channel::oneshot::Sender<Result<(), wasm_
bindgen::JsValue>>`, which does not implement the `Copy` trait
```

The compiler error is a mouthful, so it's worth breaking it down. As the error says, the `.and_then(|oneshot| Ok(oneshot.send(Ok(()))))` ; line requires the `oneshot` value to be moved into the closure. This is because `oneshot` doesn't implement copy. This makes sense; if you could copy `oneshot`, then you could use it more than once. Okay, so `oneshot` has to be moved into the closure – so what? Moves aren't bad, but the error says, `error[E0507]: cannot move out of dereference of `std::sync::MutexGuard`. Mutex takes ownership of the value you move into it, and you can't just move out its value leaving it with "nothing." So, the compiler prevents this action.

These are the kinds of errors that are both a great feature of Rust and the bane of a Rustacean's existence. Here, the compiler is preventing us from making a threading mistake, the kind that would be easy to do in almost any other language, but the side effect is an error that's hard to read. The Rust team continues to work hard on clearer compiler messages, but some things are just hard. When the compiler has you confused, read the errors slowly and carefully, and you'll usually figure out what it's trying to tell you.

So, how do you get around this problem? What you need to do is make sure to never move out of the `Mutex` reference while still getting access to the underlying `Sender`. The way we can do that is to use the `Option<T>` type, which implements a copy and the `take` function. That will allow us to replace, inside the locked `Mutex`, `Sender` with `None`. Then, any other user of that `Mutex` reference will have `None` and be able to use it appropriately.

Start by modifying the creation of `success_tx` to take `Option`, as shown in the following code:

```
let (success_tx, success_rx) =
futures::channel::oneshot::channel::<Result<(), JsValue>>();
let success_tx = Rc::new(Mutex::new(Some(success_tx)));
let error_tx = Rc::clone(&success_tx);
```

Now, in the `success` callback, we need to modify the code to account for the transmitter being optional. We'll use `take` here to immediately replace `Some(transmitter)` with `None` when its used. This is the `success` callback:

```
let callback = Closure::once(move || {
    if let Some(success_tx) = success_tx.lock().ok()
        .and_then(|mut opt| opt.take()) {
        success_tx.send(Ok(()));
    }
});
```

Here, we've used the `if let` construct to get the transmitter out of `Mutex` and `Option`. If you follow the code from `success_tx.lock()`, you'll see we call `ok` to convert the `lock()` result to an `Option`, use the `and_then` function to operate on the `Some` version of `Option`, and then finally use `take` to get the value of `Option`. In the `if` condition, we call the transmitter's `send` function with an `Ok` result, and we no longer need the strange `Ok` wrapper around the `send` call. The key is that `Option` never moves out of `Mutex`; it's replaced by `None`. Since nobody else can access the `oneshot` struct while in the lock, the code is thread-safe, and because we use `Option`, the `Mutex` always contains something – even if it's `None`.

We can finally write the `error` callback that started all this, and it's very similar:

```
let error_callback = Closure::once(move |err| {
    if let Some(error_tx) = error_tx.lock().ok()
        .and_then(|mut opt| opt.take()) {
        error_tx.send(Err(err));
    }
});
...
```

That `error` callback needs to be set using the `set_onerror` call. We had that previously, but just in case you didn't add that earlier, it looks like the following:

```
image.set_onload(Some(callback.as_ref().unchecked_ref()));
image.set_onerror(Some(error_callback.as_ref().unchecked_
ref()));
...
```

I placed the `set_onerror` call right under the existing `set_onload` call for symmetry. We do not need to add a second `await` call for the errors. Both `oneshot` transmitters send to the same receiver, because `error_tx` is a clone of `success_tx`, and we are protected from receiving an error and success because `oneshot` can only fire one time.

Now, we're handling the error and success cases correctly, and we aren't getting compiler errors. If you look at your browser right now, you should see just the triangle, as we aren't stuck at the `await` call anymore. Go ahead and restore the call to `image.set_src("Idle (1).png")` so that it uses the right file again and the RHB shows back up.

So, there it is – our game now displays an image again *and* handles errors. But what if your game displays...more than one image?

Sprite sheets

Creating a game where every sprite is its own individual file is certainly possible, but it would mean making the player wait for every file to load individually when the game started. One common way to organize sprites for a game is a **sprite sheet**, which is made up of two parts. The first is an image file with many sprites in it, like this one:

Figure 2.4 – The top of the sprite sheet

The second part is a map of coordinates and metadata that lets us "cut out" each image we need, like a cookie cutter. For instance, if we want to show the first sprite in the preceding figure (which happens to be named `Dead (7).png`), we'll need to know its location and dimensions:

Figure 2.5 – One sprite in the sheet

I've drawn a box marking the frame you'd want to "cut out" of the image when you want to draw `Dead (7).png`. When you want to draw a different file, say `Slide (1).png`, you can use the same image but a different frame when drawing.

In order to know the frame and names of every sprite sheet, we need to load a separate file that stores all that information alongside the image itself. In our case, we will use files I've already generated for you with a tool called *TexturePacker* (https://www.codeandweb.com/texturepacker), which lets you export a JSON file that looks like this:

```
{"frames": {

"Dead (1).png":
{
      "frame": {"x":0,"y":0,"w":160,"h":136},
      "rotated": false,
      "trimmed": false,
      "spriteSourceSize": {"x":0,"y":0,"w":160,"h":136},
      "sourceSize": {"w":160,"h":136}
}

...
```

TexturePacker generated a JSON file with a lookup table by sprite name. In this case, the "Dead (7).png" sprite is found at (0, 0) with a width of 109 pixels and a height of 67 pixels, so in the upper left-hand corner of the larger image. To draw the image, you'll eventually use a version of the drawImage function that takes source coordinates, which are the dimensions you saw in the preceding code, and destination coordinates where you want to position the drawing on the canvas.

So, to render the same Idle (1).png that we rendered earlier from the sprite sheet, we need to do the following:

1. Load the JSON file.
2. Parse the JSON file into a Rust structure.
3. Load the image into HtmlImageElement.
4. Use the version of drawImage that lets us draw only a part of an image element.

There isn't anything else to do, so let's get started.

Loading JSON

Inside the assets that you downloaded earlier, there's a directory called `sprite_sheets` that has two files, `rhb.json` and `rhb.png`. Go ahead and copy both of those to the `static` directory so that they can be loaded by our project. Now, let's go back and start editing `lib.rs` to load our sheet.

In this case, we'll start by writing an entirely new function to call `fetch_json`. It will use the `window.fetch` call to retrieve the JSON file and then pull the JSON off of the response body. This requires two asynchronous calls, so we're going to write the entire thing as an `async` function. Go ahead and put all of this after `main`:

```
async fn fetch_json(json_path: &str) -> Result<JsValue,
JsValue> {
    let window = web_sys::window().unwrap();
    let resp_value = wasm_bindgen_futures::JsFuture::from(
        window.fetch_with_str(json_path)).await?;
    let resp: web_sys::Response = resp_value.dyn_into()?;

    wasm_bindgen_futures::JsFuture::from(resp.json()?).await
}
```

There are a few things that won't even compile yet, and we'll fix them as I walk through this line by line.

First, we retrieve `window`. Once again, we're using `unwrap` because `window()` is an `Option`; in the next chapter, we'll do a better job of dealing with our errors. That second line is a doozy; we'll go through it in parts:

```
let resp_value = wasm_bindgen_futures::JsFuture::from(
    window.fetch_with_str(&"rhb.json")).await?;
```

The first part is the call to `wasm_bindgen_futures::JsFuture::from`, which is a little misleading. `JsFuture` is not a JavaScript future but a Rust future backed by a JavaScript promise. We want a Rust future so that we can eventually call `await` on it. We call `from` with the following:

```
window.fetch_with_str(json_path)
```

This corresponds to the `window.fetch` function in JavaScript, but as with many other JavaScript functions, `fetch` is overloaded, so we need to explicitly call it `with_str`. That function returns `Promise`, which we immediately convert to a future via the `from` call we discussed earlier. Finally, we call `await?`, which will block until `fetch` returns. This is allowed because the `fetch_json` function is `async`.

Still with me? If you understood that, you've figured out the hardest part. Next, we cast the returned `resp_value` into `Response` because the `fetch` call resolves to `JsValue`. Once again, we must convert from the dynamic typing of JavaScript to the static typing of Rust, and the `dyn_into()` function does that.

Now that we've got a response (corresponding to the `Response` object in the browser), we can call its `json()` function, corresponding to the `json()` function on the web's `Response` object. That function also returns a promise, so we wrap it in `JsFuture` as well and block on it with an `await` call.

Finally, this function returns `Result<JsValue, JsValue>`, which means it's `Result` with a dynamic JavaScript object as both its `Ok` or `Err` cases. That's why we can use `?` everywhere.

But of course, this still doesn't compile because, once again, we're missing a feature flag. Make sure you add `Response` to the list of `web-sys` dependencies, and you should be green again. Well, except for the warning that says `fetch_json` isn't called.

Parsing JSON

Back in `main`, we'll make the draw order as Red Hat Boy, the Sierpiński triangle, and then another Red Hat Boy. So, after the call to `sierpinski`, let's fetch the `"rhb.json"` file corresponding to the Red Hat Boy's data file:

```
context.draw_image_with_html_image_element(&image, 0.0, 0.0);
```

```
let json = fetch_json("rhb.json").await.unwrap();
```

This fetches the JSON but doesn't parse it into a structure we can use. We have a few options for JSON parsing, including using the browser's built-in facilities, but this is a Rust book, so let's use a Rust library, **Serde**.

Serde is one of the more popular serialization libraries for Rust and is excellent at taking JSON (as well as many other formats) and converting it to Rust structures. Add the necessary dependency to `Cargo.toml`:

```
serde = {version = "1.0.131", features = ["derive"] }
```

The crate we need is `serde`, which generically handles serialization and deserialization (**ser***/***de** – get it?). With that out of the way, go ahead and open the `rhb.json` file that you copied to the `static` directory earlier in your editor. At the top, you should see something like this:

```json
{"frames": {

"Dead (1).png":
    {
        "frame": {"x":0,"y":0,"w":160,"h":136},
        "rotated": false,
        "trimmed": false,
        "spriteSourceSize": {"x":0,"y":0,"w":160,"h":136},
        "sourceSize": {"w":160,"h":136}
    }
    ...
```

This JSON document describes a hash of frames, where the key to each frame is the name of the image (`"Dead (1).png"`) and the structure below it is the properties of that image. The property we care about is `frame`. The image for `"Dead (1).png"` is located at (`210, 493`) with a width of 71 pixels and a height of 115 pixels. Go back to the code, and we can parse that JSON that we fetched earlier.

First, we need to set up data structures that `serde` can use. At the top of `lib.rs`, we can add the `Deserialize` procedural macro to the scope:

```rust
use serde::Deserialize;
```

You'll also want to add `HashMap` from `std::collections`:

```rust
use std::collections::HashMap;
```

Now, we'll work backward. You'll have a `Sheet` class that contains the lookup table from the preceding JSON. You can put this struct anywhere in the `lib.rs` file, just not inside a function. I put it at the top:

```rust
#[derive(Deserialize)]
struct Sheet {
    frames: HashMap<String, Cell>,
}
```

The [derive(Deserialize)] macro means we can use Sheet as a target for deserializing the JSON, and HashMap and String work automatically, but we haven't defined Cell. This will represent the portion of the JSON containing frame, which is what we care about because it's where the target sprite is located. We'll add all the structs we need above Sheet:

```
#[derive(Deserialize)]
struct Rect {
    x: u16,
    y: u16,
    w: u16,
    h: u16,
}


#[derive(Deserialize)]
struct Cell {
    frame: Rect,
}
```

Great – we have a bunch of structures that can hold the map of data we need to draw our images, but we haven't filled them, but fortunately, wasm-bindgen makes this very easy with the serde-serialize feature. To enable that feature, you'll need to once again update Cargo.toml, replacing the basic wasm-bindgen dependency with the following:

```
wasm-bindgen = { version = "0.2.78", features = ["serde-
serialize"] }
```

Where before you only had wasm-bindgen = "0.2.78", now you'll need to add the serde-serialize feature flag, so you have to use the slightly more complex syntax. After that builds, you can import the JSON data with only one line of code, into_serde, after you fetch the JSON:

```
let json = fetch_json("rhb.json")
    .await
    .expect("Could not fetch rhb.json");
let sheet: Sheet = json
    .into_serde()
    .expect("Could not convert rhb.json into a Sheet
    structure");
```

I removed the unwrap calls and replaced them with expect because I wanted a specific message in these cases.

> **Tip**
> Almost all of the dependencies that we are using are very young, and it's unlikely that this book will be able to keep up with every quirk. To follow along, stick to versions the book is using, but for your own future projects, remember to check documents for feature flags, version numbers, or both whenever a dependency seemingly doesn't work.

Now that we have the sheet, we are ready to load the image and draw a sprite in it.

Drawing with our "cookie cutter"

Recall that we had four steps to draw from a sprite sheet. We've completed the first two:

1. Load the JSON file.
2. Parse the JSON file into a Rust structure.
3. Load the image into HtmlImageElement.
4. Use the version of drawImage that lets us draw only a part of an image element.

Step 3 is something you've already done before, and like all good programmers, we go immediately to one tool when we need to write the same code twice...

Copy and paste, of course! What, you thought that I was gonna say a function? We'll save that for later.

> **Tip**
> More seriously, copying and pasting to get something working a second time is perfectly acceptable; just avoid checking that in as the final version.

Copy everything from let (success_tx, success_rx) to success_rx.await and paste it right below, where we converted rhb.json into Sheet:

```
let sheet: Sheet = json
    .into_serde()
    .expect("Could not convert rhb.json into a Sheet
    structure");
```

```
let (success_tx, success_rx) =
futures::channel::oneshot::channel::<()>();
```

```
...
```

```
image.set_src("Idle (1).png");
success_rx.await;
```

Thanks to the way Rust works, you won't need to rename any variables, as every time you use `let`, you shadow the previous version of that variable and create a new binding. In the pasted code, we only need to make one change – to load the image sheet instead of `"Idle (1().png"`:

```
image.set_src("rhb.png");
```

Step 3 is now complete; we've loaded the large image with many sprites in it. Finally, we'll draw the sprite that we want. Let's go ahead and draw the `"Run (1).png"` sprite, which will admittedly look similar but will allow us to add some animation to go along with it. We'll use the version of the `drawImage` call that takes a source location, which is the frame we discussed earlier, and a destination location where we will put the image on the canvas. To make sure that we see the new image, let's stick this image somewhere near the middle. Add this right after the last `await` call:

```
let sprite = sheet.frames.get("Run (1).png").expect("Cell not
found");
context.draw_image_with_html_image_element_and_sw_and_sh_and_
dx_and_dy_and_dw_and_dh(
    &image,
    sprite.frame.x.into(),
    sprite.frame.y.into(),
    sprite.frame.w.into(),
    sprite.frame.h.into(),
    300.0,
    300.0,
    sprite.frame.w.into(),
    sprite.frame.h.into(),
);
```

The first line, sheet.frames.get, retrieves the sprite by name, with an expect thrown in for when we get the name wrong. The next line is a monster because drawImage has nine argument versions in JavaScript, and it's represented in Rust by the call to draw_image_with_html_image_element_and_sw_and_sh_and_dx_and_dy_and_dw_and_dh. That's a mouthful, but what it means is drawing the image using the source rectangle (our frame) to a destination rectangle, where the source rectangle is represented by four position and size coordinates, and the destination rectangle is also represented by four coordinates. The source rectangle is our frame, drawn from the JSON file we loaded earlier. The destination rectangle starts at (300,300) to put RHB in about the center of the canvas and uses the same width and height because we don't want to change the size of the image. The end result is this:

Figure 2.6 – Multiple Red Hat Boys

The original RHB is up in the left, using its own image file, and the second RHB from the sprite sheet is approximately in the center of the triangle. You'll notice his right hand is slightly tucked in because it's at the start of his run animation.

Speaking of the run animation, how about we see it in action?

> **Important Note**
>
> Loading the sprite sheet and images the way we did here is just one of many ways to implement this technique. For example, another option would have been to embed the JSON and the images in the Rust executable, perhaps by Base64-encoding them, thereby doing all data loading at once. They could also have been bundled into destination applications via webpack and exposed to our Rust app. All these different ways come with their own trade-offs, and in our case, we have traded some complexity and upfront load times for the requirement to make calls to a server. Find the solution that works best for your game.

Adding animation

Sprite animation works just like a flip-book or a movie. Show a sequence of images fast enough, where each image is drawn to be only slightly different than the previous one, and it causes the illusion of motion. Animation on the canvas works in much the same way, where each frame in the sprite sheet has the same effect as a drawing in a flip-book:

Figure 2.7 – The run animation for Red Hat Boy

To draw Red Hat Boy running, we have to simply draw the images in order, one at a time, and loop after drawing the last one. Simple for a loop, right?

Of course, it's not quite that simple. First, we can't just use an infinite loop, as that would block the browser from any processing, resulting in a frozen browser tab. Second, we have to make sure to clear the canvas between each frame. Otherwise, we'll see all the images merged together as one draws on top of the other. So, each time we draw the canvas, we'll need to clear it first, and then draw the desired frame.

> **Important Note**
>
> If you're familiar with double buffering in traditional game development and are worried about seeing flicker when we clear the canvas and then redraw, don't. The canvas element already handles this for you.

Fortunately, you already know almost all you need to in order to draw the animated RHB. You'll need to pass a Rust closure to a function and draw a sprite from a sprite sheet. The only thing you don't know is how to clear the canvas, and we'll cover that in a moment, but we must start by saying goodbye:

1. **Delete the Sierpiński triangle**: We could try and draw the Sierpiński triangle over and over again after clearing the screen, but it would be distracting and slow. So, it's time to say goodbye, to the call to `sierpinski` and all the code it uses, including the `midpoint` and `draw_triangle` functions. They served us well and will be missed.

2. **Delete the idle RHB**: We could probably go to the effort of keeping the idle RHB sprite around, but it would require dealing with the duplicated code we wrote to create the sprite sheet. It's best to delete all that copy and paste code before the boss finds out.

    No, go ahead and delete everything inside the `spawn_local` closure up until we loaded the `rhb.json` file. After those deletions, your code should look like this around `spawn_local`:

```
let context = canvas
    .get_context("2d")
    .unwrap()
    .unwrap()
    .dyn_into::<web_sys::CanvasRenderingContext2d>()
    .unwrap();

wasm_bindgen_futures::spawn_local(async move {
    let json = fetch_json("rhb.json")
        .await
        .expect("Could not fetch rhb.json");
    ...
```

So, before spawning the local future, the last thing you do is get the 2d context, and the first thing you do after spawning the future is load the JSON.

Now, it's time to change the draw into a callback function.

3. **Draw on SetInterval**: We want to draw our running RHB at approximately 20 frames per second, so we'll need to make the draw call every 50 milliseconds. You might be wondering why we don't have the animation go at 60 frames per second, the traditional frame rate of a game, and that's because the sprite itself isn't designed to run that fast. He'll look like a sped-up version of Usain Bolt. In the next chapter, when we set up a proper game loop, we'll run at 60 frames per second and only update the sprite every third tick to compensate, but for now, let's set up a callback every 50 milliseconds with JavaScript's `setInterval` function, which is called `set_interval_with_callback`. First, we need to set up the callback itself, using the `Closure` struct that we used earlier. Right after the `success_rx.await` call, add this:

```
let interval_callback = Closure::wrap(Box::new(move ||
{})) as Box<dyn FnMut()>);
```

This sets up an empty `Closure`, but unlike the previous time we created `Closure`, we're using `Closure::wrap` instead of `Closure::once`. Why? Well, because this closure will be called multiple times. This also means we need to use `Box` with an explicit cast, as `Box<dyn FnMut()>`, because the `wrap` function requires `Box`, and there isn't enough information for the compiler to infer the type.

Now that we have an empty interval callback, we can schedule it to be called. On the next line, add the following:

```
window.set_interval_with_callback_and_timeout_and_arguments_0(
    interval_callback.as_ref().unchecked_ref(),
    50,
);
```

Adding that will start the process of calling our `interval_callback` every 50 milliseconds; however, doing so will cause an error. If you look into the browser's error log via the console, you'll see this repeated:

```
Uncaught Error: closure invoked recursively or destroyed
already
```

That should sound familiar, as we've already fixed it once this chapter. The fix will be to once again forget the closure that we passed into `setInterval` so that Rust doesn't destroy it when we leave the scope of this future. Add this right *after* the `set_interval` call:

```
interval_callback.forget();
```

Then, go back and check the console to verify that the error has gone away. You may need to refresh the browser to ensure that you don't get stale error messages showing up to confuse you.

Now that you've scheduled a regular callback, let's add one line to that callback to clear the screen:

```
let interval_callback = Closure::wrap(Box::new(move || {
    context.clear_rect(0.0, 0.0, 600.0, 600.0);
}) as Box<dyn FnMut()>);
```

This will not compile because outside of this callback, we're still calling `draw_image`. Since we've moved the `context` into this `Closure`, we've run afoul of the borrow checker. To address this, we're going to need to move the drawing code into the closure, like so:

```
let interval_callback = Closure::wrap(Box::new(move || {
    context.clear_rect(0.0, 0.0, 600.0, 600.0);
    let sprite = sheet.frames.get("Run(1).png").expect
        ("Cell not found");
    context.draw_image_with_html_image_element_and_sw_and_sh_
and_dx_and_dy_and_dw_and_dh(
        &image,
        sprite.frame.x.into(),
        sprite.frame.y.into(),
        sprite.frame.w.into(),
        sprite.frame.h.into(),
        300.0,
        300.0,
        sprite.frame.w.into(),
        sprite.frame.h.into(),
    );
}) as Box<dyn FnMut()>);
```

Congratulations! You are now clearing the screen and redrawing it every 50 milliseconds. Unfortunately, it doesn't look like anything because you're always drawing the same image. Let's change the code to loop from "Run (1).png" to "Run (8).png" over and over again.

Initialize a frame counter outside of the closure:

```
let mut frame = -1;
let interval_callback = Closure::wrap(Box::new(move || {
```

Now, on the inside of the closure, we'll cycle the frame count between 0 and 7:

```
let interval_callback = Closure::wrap(Box::new(move || {
    frame = (frame + 1) % 8;
```

Why 0 to 7 when it goes to frame 8? Because we'll adjust it on the next line when we construct framename:

```
let interval_callback = Closure::wrap(Box::new(move || {
    frame = (frame + 1) % 8;
    let frame_name = format!("Run ({}).png", frame + 1);
```

Finally, instead of getting "Run (1).png" every time, we'll get the constructed sprite name from the sprite sheet. Just change the sheet.get call to use &frame_name, and we'll move the call to get above the clear_rect call as well:

```
let frame_name = format!("Run ({}).png", frame + 1);
let sprite = sheet.frames.get(&frame_name).expect("Cell not
found");
```

Take a look now and, sure enough, Red Hat Boy is running!

Figure 2.8 – You can't see running in a book, trust me

Summary

In this chapter, we have covered rendering sprites to the screen, including sprite sheets, but we actually covered so much more than that. We covered how to use futures and `async` code in a WebAssembly app, how to parse JSON, and perhaps most confusingly how to send Rust closures to JavaScript via the `Closure` struct. We also reviewed some of the quirks of using Rust in the WebAssembly environment from *Chapter 1, Hello WebAssembly*. This chapter was fun, but we made some messy code.

In the next chapter, we'll deal with that by setting up a simple architecture for our game and writing a proper game loop. Lest you think *Chapter 3, Creating a Game Loop*, is all refactoring, we'll also move our friend Red Hat Boy around the screen. It'll start to look like a real game!

Part 2: Writing Your Endless Runner

Now that you've got the skeleton of a Rust and WebAssembly book put together, you'll write a full-fledged game using this unique toolchain. The game, an endless runner called "Walk the Dog," will need user input, a game loop, sounds, and more. By the end of this section, you'll have your own version running that you can play repeatedly, with its web-based UI.

In this part, we cover the following chapters:

3
Creating a Game Loop

In the first two chapters, we focused on getting an application built, an environment set up, and graphics on a screen without concerning ourselves with creating an actual functioning game. There's no interactivity here, and no straightforward way to add more characters without copying and pasting more code. In this chapter, that will change, with the addition of a game loop and keyboard events, but first, we're going to need to restructure the code to make it ready for our new features. Be prepared to dig in – this is going to be a busy chapter.

We're going to cover the following:

- Minimal architecture for games
- Creating a game loop
- Adding a keyboard input
- Moving Red Hat Boy

By the end of the chapter, we'll have a mini-game engine that's ready to be extended with new features and process input.

Technical requirements

There are no new technical requirements for this chapter; I recommend making sure that your editor/IDE setup is comfortable for you. You're going to be making a lot of changes and you'll want your editor to help you along. The source code for this chapter is available at `https://github.com/PacktPublishing/Game-Development-with-Rust-and-WebAssembly/tree/chapter_3`.

Check out the following video to see the Code in Action: `https://bit.ly/3qP5NMa`

Minimal architecture

A few years ago, I had a realization while preparing a talk on HTML5 game development. The day before I was scheduled to give the talk, I had written the slides and prepared my delivery, but I had one small problem – I had no demo! I needed a demo of a game to finish off my talk; indeed, I had referenced it in my slides, so I had to produce it. If you've ever been up against a deadline, you know what happens next. All of my ideas about clean code and software architecture were thrown to the side, as I hacked and slashed my way to a working prototype of *Asteroids* in HTML5. You can still find it on my GitHub here: `https://github.com/paytonrules/Boberoids`, complete with a name that doesn't make sense.

The code, by virtually any standard, is pretty terrible. In much the same way the code in *Chapter 1, Hello WebAssembly*, and *Chapter 2, Drawing Sprites*, proceeds in a straight line with no modules, separation of concerns, or tests, this code brute-forces its way from the start to the end of the program. But a funny thing happened at about 2 AM the day before that presentation – it worked! In fact, in preparation for this chapter, I cloned the nearly 10-year-old program, ran `python -m http.server`, browsed to `http://localhost:8000`, and, well, here you go – a mostly working clone of *Asteroids*:

Figure 3.1 – Asteroids with a company logo

Of course, this code is also nearly impossible to extend or debug. There's no encapsulation and it's all in one file; heck, there isn't even a proper README file. By any objective measure, this is **bad** software.

It was so bad that while working on this presentation, I began simultaneously working on an open source project called *"Eskimo"* (https://github.com/paytonrules/Eskimo), which was meant to be a **good** game framework with the best object-oriented design I knew at the time, with a test-first approach and things like CI built in. If you look at the commit dates, you may notice that my last commit on this project was 2 years after the *Asteroids* clone you can see in the preceding screenshot. The code, if you were simply doing a code review, is far better than the code for the aforementioned game. It just doesn't actually work.

I never made a working game with *Eskimo*. Like many developers before me, I fell into the trap of writing a framework instead of making a game and spent so much time "perfecting" my framework that I lost interest in the game I was allegedly making. This bothered me for a long time, and I kept asking myself the question, "Why did I finish the game when I did everything wrong, and fail when I did it right?" Does good code have any real-life meaning?

I won't keep you in suspense; for the purposes of this book, we're going to define minimal architecture as one that **makes the next feature easier**. That means that we are going to do some architecture but only enough to make things easier going forward. We'll be on the lookout for extra complexity and "gold-plating." We're making a game, not an engine, and we want to finish the game.

Good? Bad? I'm the guy with code

Minimal architecture sounds simple but can be hard, so let me explain with a counter-example.

Eskimo has an Events object that is created with a constructor that takes jquery, the document, a game object, and a canvas. It has all of this because I took the principle of dependency injection to an extreme and tried to make sure the Events object would not depend directly on any of those things.

The problem? Three of those objects are never going to change. You're not going to replace jquery, the document, or the canvas in any game, at least not with *Eskimo*, and it requires a lot of understanding to follow the Eskimo code because of that. While the code is theoretically more flexible and follows the dependency inversion principle (http://bit.ly/3uh7fWU), it actually made it harder to add future features because I couldn't keep in my head what dependencies did what. My mistake was that I injected these dependencies before I had a reason to, out of a misplaced sense of "good code."

We are going to stay focused on our goal of making a game and do not want to get caught up in making a framework. That means our process for evolving our program into a game is going to introduce a little bit of flexibility each time we need it. Returning to the two example games, *Asteroids* and *Eskimo*, we can think of them as being on a scale of rigidity. The *Asteroids* clone is extremely rigid. It's like a steel pole, and if you want to change it, you can't. You can only break it. Meanwhile, the *Eskimo* game framework is infinitely flexible, so much so that it can't actually do anything. It collapses in on itself in to a lump of goo. Our game, which is just Red Hat Boy running so far, is also very rigid. Adding a second object, say the dog, would require a lot of code changing throughout the small application and would potentially introduce defects.

So in order to take our game and add more features, particularly interactivity, we'll need to introduce some flexibility. We'll heat up our steel pole so that it can bend, bend it, and then let it harden again.

Layered architecture

We're going to start by introducing a small layered architecture. Specifically, we'll have three layers:

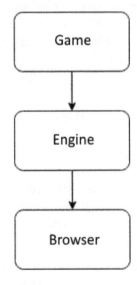

Figure 3.2 – A layered architecture

The one rule of this architecture is that layers can only use things at or below their layer. So working from the bottom, the browser layer is going to be a bunch of small functions that are specific to the browser. For instance, our `window` function will end up here. Meanwhile, the engine layer is going to be tools that work across our game, such as the `GameLoop` structure. Finally, the game is the layer that contains our actual game logic. Eventually, we'll spend most of our development time in this layer, although initially, we'll spend a lot of time in the `Engine` and `Browser` layers until they have settled.

Why do this? Our aforementioned rule was that any change in architecture has to make future changes easier, so let's identify what makes changes hard right now:

- Keeping everything in one long function makes the code hard to follow.

- Extracting all the `Browser` code will allow us to unify error handling.

The first point reflects that our brains can only hold so much. Keeping all the code in one place means scrolling up and down trying to find where things are and trying to remember virtually all of the code. Extracting code into various constructs such as modules, functions, and structs with **names** lets us reduce the amount of information in our heads. This is why the right design feels good to program in. Too much abstraction and you've replaced keeping track of all the details of the program with keeping track of all the abstractions. We'll do our best to keep things in the sweet spot.

The second reason for the layered approach is specific to Rust and the `wasm-bindgen` functions, which all return `JsValue` as their error type. While this works in a browser, it does not work well when intermingling with the rest of a Rust program because `JsValue` does not implement the `std::Error::error` type that most other Rust errors implement. That means you can't write a function like the following:

```
async fn doesnt_compile() -> Result<(), Box<dyn
std::error::Error>> {
    let window = web_sys::window()?;
    let json = fetch_json("rhb.json").await?;



    ...
}
```

The preceding code won't compile because while `ThreadPool::new` returns a `Result<ThreadPool, Error>`, `fetch_json` returns `Result<JsValue, JsValue>`, and those results don't mix. In the `browser` module, we'll map JsValues to a standard error, using the `anyhow` crate. We'll also use it to hide the weird details of the API, creating one that's tailored to our purposes. Let's get started creating our `browser` module.

Creating the browser module

The first step is to create a file named `browser.rs` in the `src` directory and reference it at the top of `lib.rs` with `mod browser`. While we could, theoretically, put every module in `lib.rs`, we're not monsters, and we'll break things into their own files. By the end of this chapter, `lib.rs` will be very small. Our first addition to `browser` is actually going to be a macro, and a completely new one, listed as follows:

```
macro_rules! log {
    ( $( $t:tt )* ) => {
        web_sys::console::log_1(&format!( $( $t )*
            ).into());
    }
}
```

I'd love to claim that I was a great macro programmer who wrote that in one try, but the truth is that that little macro is straight from the Rust and WebAssembly docs (https://bit.ly/3abbdJ9). It's a macro that allows you to log in to the console with `log!` using a syntax such as the `format!` function. In `lib.rs`, add an annotation to the `browser` module declaration, like so:

```
#[macro_use]
mod browser;
```

This makes `log!` available whenever the `browser` module is used. Given that we'll be making a lot of changes, we'll probably want some easy debugging. The next step will be to add the `anyhow` crate, which we'll use to unify the error handling across WebAssembly and pure Rust code. The dependency is added to `Cargo.toml` as `anyhow = "1.0.51"`. This crate provides a few features that we'll be using extensively:

- An `anyhow::Error` type that conforms to the `std::error::Error` trait
- An `anyhow!` macro that lets us create error messages that conform to the type, with strings
- An `anyhow::Result<T>` type that is a shortcut for `Result<T, anyhow::Error>`

Go ahead and add `use anyhow::{anyhow, Result};` to the top of the use declarations now so that we'll have them ready as we create new functions here.

Now that the `browser` module is prepared, let's work from the top of `main` and work downward, extracting functions. Let's start here:

```
#[wasm_bindgen(start)]
pub fn main_js() -> Result<(), JsValue> {
    console_error_panic_hook::set_once();

    let window = web_sys::window().unwrap();
    let document = window.document().unwrap();
```

The `wasm_bindgen` macro has to stay there, and it is only compatible with a function that returns `Result<(), JsValue>`. This means that while we can use proper Rust errors throughout our program, at the end, we'll need to transform it back to a `JsValue` if we want it to be returned from this function. Fortunately, once we write our game loop, this won't be a problem.

> **Important Note**
>
> `Wasm_bindgen` error handling is a little rough, and the Wasm working group is aware of it. For reference, you can look at the defect here: `https://bit.ly/3d8x0D7`.

Moving on to the executing code, there are two functions right off the top that can be pulled into `browser.rs`. We'll take the refactoring step by step. First, let's create a function in the browser module, like this:

```
pub fn window() -> Result<Window> {}
```

That won't compile because it doesn't return anything, but also because it doesn't know anything about the `Window` type. Go ahead and import those at the top of the file. It should look like this:

```
use web_sys::Window;
```

> **Tip**
>
> If you haven't already, get Rust Analyzer working with your editor of choice. I use emacs and use the keyboard shortcut , c a to import modules. It's a great timesaver for this kind of work. From this point forward, I won't be documenting every `use` declaration you need when moving files over; just follow the compiler errors.

The function also doesn't compile because you're not returning anything. You can start by directly copying (don't cut) the call to `window()` from `lib.rs`:

```
pub fn window() -> Result<Window> {
    web_sys::window().unwrap()
}
```

You don't need to bind a variable with `let` here. This still won't compile because of that `unwrap`. In this case, `web_sys::window` returns `Option<Window>` and `unwrap` will extract the `Window` object, or panic. None of that conforms to `Result<Window>`, and what we need to do instead is handle the case where `window` is somehow missing as an error.

> **Important Note**
>
> When trying to mix and match `Option` and `Result`, there are two schools of thought – make `Result` into `Option` with ok, or convert `Result` into `Option` with `ok_or_else`. I prefer the second because while that will mean writing a lot of error messages that say "<X>" (not found), the alternative is losing useful error diagnostics.

To make this function work with the `Result<Window>` return type, which, remember, is a shorthand for `Result<Window, anyhow::Error>`, we're going to use the `anyhow!` macro. So, to convert `Option` to `Result` and make this function compile, you can do the following:

```
pub fn window() -> Result<Window> {
    web_sys::window().ok_or_else(|| anyhow!("No Window Found"))
}
```

Now you've got a function, `browser::window()`, which will return `Window` or the appropriate error.

> **Important Note**
>
> Nightly Rust currently has an error called `NoneError` that helps bridge the gap between `Option` and `Result` types, but we'll stick to the standard for now.

Finally, we can replace the call to web_sys::window() in lib with a call to browser::window() in lib:

```
let window = browser::window().expect("No Window Found");
let document = window.document().unwrap();
```

The call to window() will use expect for now to go ahead and crash the program if there is no window. Later, you'll see we can use the ? operator, but for the moment, we've got to work around main_js returning a Result<(), JsValue>. If that was the only place we were changing, the introduction of anyhow wouldn't make any sense. Fortunately, when we repeat that process with a new document function in the browser module, you can see the advantage. We can skip going through each step of that process, and get to the end result:

```
pub fn document() -> Result<Document> {
    window()?.document().ok_or_else(|| anyhow!
        ("No Document Found"))
}
```

If this doesn't compile, don't forget to add Document to the use declarations at the top of the module. As we make these changes, you'll need to move use declarations into browser, but you'll be able to remove them from lib.rs.

You can now actually shrink the two calls to window() and document() in lib.rs into one call, like so:

```
pub fn main_js() -> Result<(), JsValue> {
    console_error_panic_hook::set_once();

    let document = browser::document().expect("No Document Found");
    ...
```

There is one place in lib.rs where we're using the window variable we just deleted. Near the bottom of spawn_local Closure, right after creating interval_callback, there is a call to window.set_interval_with_callback_and_timeout_and_arguments_0 that can replace window with browser::window().unwrap(). That looks like the following:

```
let interval_callback = Closure::wrap(Box::new(move || {
    ...
}) as Box<dyn FnMut()>);
```

```
browser::window()
    .unwrap()
    .set_interval_with_callback_and_timeout_and_arguments_0(
        interval_callback.as_ref().unchecked_ref(),
        50,
    );
interval_callback.forget();
```

Our next function will get the canvas object, but it's a little more complicated than the previous two functions. We were pretty casual with the unwrap calls for that section, so we'll have to do some converting to get more specific errors. The end result looks like this:

```
pub fn canvas() -> Result<HtmlCanvasElement> {
    document()?
        .get_element_by_id("canvas")
        .ok_or_else(|| anyhow!
            ("No Canvas Element found with ID 'canvas'"))?
        .dyn_into::<web_sys::HtmlCanvasElement>()
        .map_err(|element| anyhow!("Error converting {:#?}
            to HtmlCanvasElement", element))
}
```

There are a few things worth paying close attention to here. First, the get_element_by_id call is hardcoded to the 'canvas' ID. We'll go ahead and leave that as is until it causes an issue later, but we're not going to make that configurable until we need to. Next, we used ok_or_else to convert get_element_by_id from Option to Result. Most interesting is the call to the dyn_into function. As discussed earlier, almost every function that calls into JavaScript will return a JsValue type, because JavaScript is a dynamically typed language. We know that the element returned by get_element_by_id will return HtmlCanvasElement, at least if we've retrieved the right JavaScript node, so we can convert it from JsValue to the correct element. This is what dyn_into does – it converts from JsValue to appropriate Rust types. In order to use dyn_into, you must import wasm_bindgen::JsCast, which rust-analyzer cannot automatically import. It can import web_sys::HtmlCanvasElement.

We'll create a `context` function that looks very similar:

```
pub fn context() -> Result<CanvasRenderingContext2d> {
    canvas()?
        .get_context("2d")
        .map_err(|js_value| anyhow!("Error getting 2d
          context {:#?}", js_value))?
        .ok_or_else(|| anyhow!("No 2d context found"))?
        .dyn_into::<web_sys::CanvasRenderingContext2d>()
        .map_err(|element| {
            anyhow!( "Error converting {:#?} to
                     CanvasRenderingContext2d",
                     element
            )
        })
}
```

One oddity you might see here is that we follow `map_err` immediately with `ok_or`.
That's because `get_context` returns `Result<Option<Object>, JsValue>`,
which the old code "solved" by calling `unwrap` twice. So what we do now is map the error
(`JsValue`) to `Error` and then take the inner `Option` and map the `None` case to a value.

Remember that if you're following along and having trouble compiling, update your `use`
declarations. Let's pick up the pace a little. We can add a function for `spawn_local`:

```
pub fn spawn_local<F>(future: F)
where
    F: Future<Output = ()> + 'static,
{
    wasm_bindgen_futures::spawn_local(future);
}
```

> **Tip**
> If you are writing a wrapper like this and aren't sure what the signature should
> be, start by looking at the function you're wrapping and mimic its signature.

Let's also add fetching JSON to browser:

```
pub async fn fetch_with_str(resource: &str) ->
  Result<JsValue> {
    JsFuture::from(window()?.fetch_with_str(resource))
        .await
        .map_err(|err| anyhow!("error fetching {:#?}",
                               err))
}
```

```
pub async fn fetch_json(json_path: &str) -> Result<JsValue> {
    let resp_value = fetch_with_str(json_path).await?;
    let resp: Response = resp_value
        .dyn_into()
        .map_err(|element| anyhow!("Error converting {:#?}
            to Response", element))?;

    JsFuture::from(
        resp.json()
            .map_err(|err| anyhow!("Could not get JSON from
                response {:#?}", err))?,
    )
        .await
        .map_err(|err| anyhow!("error fetching JSON {:#?}", err))
}
```

I expanded fetch_json into two functions because I think fetch_with_str is going to be reusable, but it's not strictly necessary. The fetch_json function borders on not belonging in the browser module. On the one hand, it exclusively calls into the wasm_bindgen API, mapping the JsValue errors to standard Errors; on the other hand, there is a tiny amount of behavior there when we decide to get JSON off of the response. Ultimately, that's a bit of a judgment call.

Having written all those functions, you can go back to the lib.rs module and update the main function to use the new ones. As you can see, it's starting to shrink significantly, as the top should look like the following, using the new functions from the browser module where appropriate:

```
#[wasm_bindgen(start)]
pub fn main_js() -> Result<(), JsValue> {
    console_error_panic_hook::set_once();
    let context = browser::context().expect("Could not get
      browser context");

    browser::spawn_local(async move {
        let sheet: Sheet = browser::fetch_json("rhb.json")
            .await
            .expect("Could not fetch rhb.json")
            .into_serde()
            .expect("Could not convert rhb.json into a
                    Sheet structure");

        let image =
            web_sys::HtmlImageElement::new().unwrap();
...
```

You can see that we removed all the intermediate calls to window and context in favor of one call to context. We've also just made a call into fetch_json using expect to call out errors. Finally, you'll see one compiler error when you do this on the window. set_interval_with_callback_and_timeout_and_arguments_0 line. You can fix that by replacing window with browser::window().unwrap(). The unwrap bit is ugly, but we'll keep refactoring until that's gone as well. It's not reproduced in the preceding snippet, but you can also delete the fetch_json function from lib.rs; it's not being used anymore.

This brings us to the next section to extract – loading an image.

Loading an image

I bet you thought you were done with loading the image, didn't you? Well, we will be as soon as we turn it into a function. Let's look at the original implementation for a moment again:

```
let image = web_sys::HtmlImageElement::new().unwrap();
let (success_tx, success_rx) =
    futures::channel::oneshot::channel::<Result<(),JsValue>>();
let success_tx = Rc::new(Mutex::new(Some(success_tx)));
let error_tx = Rc::clone(&success_tx);

let callback = Closure::once(Box::new(move || {
    if let Some(success_tx) =
        success_tx.lock().ok().and_then(|mut opt| opt.take())
    {
        success_tx.send(Ok(()));
    }
}));

let error_callback = Closure::once(Box::new(move |err| {
    if let Some(error_tx) =
        error_tx.lock().ok().and_then(|mut opt| opt.take()) {
        error_tx.send(Err(err));
    }
}));

image.set_onload(Some(callback.as_ref().unchecked_ref()));
image.set_onload(Some(error_callback.as_ref().unchecked_ref()));
image.set_src("rhb.png");

success_rx.await;
```

At first glance, this looks like it's one function in our `browser` module, `load_image`, but on closer reflection, there's a lot here for just one function. For instance, if you so choose, you can create an image element without worrying about whether it's going to be loaded, or you might be willing to use `set_src` without concerning yourself with whether or not it's loaded. No, all that stuff after `let image = web_sys::HtmlImageElement::new().unwrap()` is really engine behavior. That means it's time for us to create our second module, `engine`!

The `engine` module will contain libraries and functions that we will use throughout our game. We are **not** going to write a fully fledged commercial-quality engine in this book because we want to finish an actual game, but we are going to have engine-like functions and structures, and we'll put those in the `engine` module. In fact, to break down this behavior, we'll follow a few steps:

1. Create a `browser` function, `new_image`.
2. Create a `browser` function to create JS closures.
3. Create an `engine` module.
4. Create an `engine` function, `load_image`.

Let's start with the changes to `browser`; we'll create two new functions to make `Closure` and an image:

```
pub fn new_image() -> Result<HtmlImageElement> {
    HtmlImageElement::new().map_err(|err| anyhow!("Could
        not create HtmlImageElement: {:#?}", err))
}


pub fn closure_once<F, A, R>(fn_once: F) ->
  Closure<F::FnMut>
where
      F: 'static + WasmClosureFnOnce<A, R>,
{
    Closure::once(fn_once)
}
```

The first function is just a wrapper around `HtmlImageElement`; there's not much to explain. In the future, we may decide we want our own type for images, but for now, we'll stick with the browser-provided type. The `closure_once` function is complicated by its type signature. In this case, we just mimic the exact same type signature of the `Closure::once` function from `wasm_bindgen`. Later, we'll write some utility functions for the `Closure` types to make working with them easier, but for this one, we'll just create a straight wrapper.

> **Important Note**
>
> A compelling argument can be made that we should be converting even more types in this module. Specifically, we should use our own types for `Closure`, `HtmlImageElement`, and other browser-provided types. It's possible that's a better approach, but for now, we're going to stick with the types provided in the interest of both learning the material and keeping to a simple architecture. Like all decisions in programming, it's a trade-off.

That covers *step 1* and *step 2*, and *step 3* is quick – create a file named `engine.rs` in the source directory and add a `mod engine` declaration to `lib.rs`. Now for *step 4*, the one we've been dreading. In `engine.rs`, add the following:

```rust
pub async fn load_image(source: &str) ->
Result<HtmlImageElement> {
    let image = browser::new_image()?;

    let (complete_tx, complete_rx) =
      channel::<Result<()>>();
    let success_tx =
      Rc::new(Mutex::new(Some(complete_tx)));
    let error_tx = Rc::clone(&success_tx);
    let success_callback = browser::closure_once(move || {
        if let Some(success_tx) =
            success_tx.lock().ok().and_then(
            |mut opt| opt.take()) {
            success_tx.send(Ok(()));
        }
    });

    let error_callback: Closure<dyn FnMut(JsValue)> =
```

```
    browser::closure_once(move |err| {
        if let Some(error_tx) =
            error_tx.lock().ok().and_then(
            |mut opt| opt.take()) {
                error_tx.send(Err(anyhow!("Error Loading Image:
                    {:#?}", err)));
        }
    });

    image.set_onload(Some(
        success_callback.as_ref().unchecked_ref()));
    image.set_onerror(Some(
        error_callback.as_ref().unchecked_ref()));
    image.set_src(source);

    complete_rx.await??;

    Ok(image)
}
```

I'm intentionally leaving out the use statements so that you get used to adding them and thinking about which declarations you need and are using. However, there are two traps to this code that I want to call out:

- In order for unchecked_ref to compile, you need to use wasm_bindgen:JsCast.

- When you import channel, make sure you choose futures::channel::oneshot::channel. There are a few different implementations of channel, and if you grab the wrong one by mistake, this code won't compile.

When in doubt, take a look at lib.rs and verify which dependencies are being used there because that's where this code is being pulled from.

Returning to the code we added, note that we're using our new `browser` functions throughout, with no direct dependencies on the `wasm-bindgen` functions. We are still dependent on `wasm_bindgen` for the `Closure` and `JSValue` types, as well as the `unchecked_ref` function, but we've reduced the amount of direct platform dependencies. Our only JS dependency is on `HtmlImageElement`. Now, take a look at the very beginning of the function and you'll see the `new_image` call can use the `?` operator to early return in the event of an error, with a standard Rust error type. This is why we mapped those errors in the `browser` functions.

Moving past the first two lines of the method, the rest of the function is largely the same as before, replacing any direct calls to `wasm-bindgen` functions with their corresponding calls in `browser`. We've changed the channel to send `anyhow::Result` and used `anyhow!` in `error_callback`. This then allows us to end the function with a call to `complete_rx.await??` and `Ok(image)`. Those two `??` are not a misprint; `complete_rx.await` returns `Result<Result<(), anyhow::Error>, Canceled>`. Since `anyhow::Error` and `Canceled` both conform to `std::error::Error`, we can handle those errors with `?` each time.

We still have two warnings in this function because both of the calls to `send` return `Result` that we aren't dealing with. We can't just use `?` because those results are wrapped in the `Closure` types, so we'll put off dealing with those unlikely errors for now and will cover error logging in *Chapter 9*, *Testing, Debugging, and Performance*.

Now that you've done all that you should be able to, replace the code in `main` with a call to our new function:

```
let sheet: Sheet = json
    .into_serde()
    .expect("Could not convert rhb.json into a Sheet
        structure");

let image = engine::load_image("rhb.png")
    .await
    .expect("Could not load rhb.png");

let mut frame = -1;
```

Nothing about loading `Sheet` has changed; that's just there to make sure you put this in the right place. After that, the code for animating our little **Red Hat Boy** (**RHB**) starts, but we're not going to be using that at all. That will be replaced with our game loop, which we'll start introducing now.

Creating a game loop

The core of this game, and virtually every game ever, is just an infinite loop. You can boil them all down to something like this:

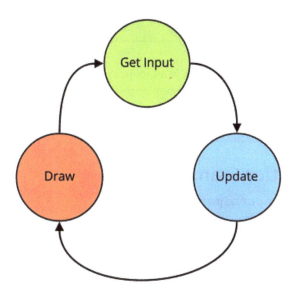

Figure 3.3 – A basic game loop

That means, theoretically, these are trivial to implement, as shown here:

```
while(!quit) {
    handleInput()
    updateGame()
    drawGame()
}
```

At its core, that's what we're going to write, but as you probably guessed, if it was that simple, I wouldn't have devoted an entire chapter to it. No, there are two problems we are going to be dealing with as we write it:

- **The browser**: If we were writing this game as a command-line program, we'd be able to use the preceding loop, but not in the browser. Any program running in the browser **must** give up control to the browser itself so that it can do whatever a browser does when it's not showing cat videos, and this kind of loop would cause the current browser tab to hang. It's a big no-no, and to get around it, we need to use the requestAnimationFrame function.

- **Frame rates and physics**: The preceding loop we wrote would run as fast as the computer could go. Well, is every computer on the internet the same speed? No, of course not, so we'll need to make sure we can account for the differences between machine speeds in our loop, as best we can. We'll do that with what's called a **fixed-step** game loop.

> **Important Note**
>
> You could probably write a book on game loops if you so chose, but this section owes a heavy debt to `https://gameprogrammingpatterns.com/game-loop.html` and `https://gafferongames.com/post/fix_your_timestep/`.

RequestAnimationFrame

We'll start with the `requestAnimationFrame` function, which is a browser function that "requests" a new frame draw as soon as possible. The browser then fits that in frame draw in between handling things such as mouse clicks, operating system events, and cat videos. You might think that would be very slow, but in fact, typically, it's able to render at 60 frames per second, provided your game can keep up. The catch is that unlike our `setInterval` call from earlier, this function needs to be called at the end of every animation. A fairly straightforward JavaScript version of an animation might look like this:

```
function animate(now) {
    draw(now);
    requestAnimationFrame(animate);
}

requestAnimationFrame(animate);
```

So, `requestAnimationFrame` is called with the `animate` function as its parameter. The browser then calls the `animate` function on the next frame, which draws and then requests the next frame. This looks like an infinite loop, but, in fact, doesn't block the browser because of the calls to `requestAnimationFrame`. This also takes a parameter, `now`, which is a timestamp in milliseconds of when the callback function was started. We'll use this to fix our physics as we evolve our game loop, but writing a game loop in Rust is a little weird because of the borrowing guarantees, so let's start by writing a very basic one.

You can start by adding a simple wrapper for `requestAnimationFrame` to `browser`, as shown in the following code:

```
pub fn request_animation_frame(callback: &Function) ->
  Result<i32> {
    window()?
         .request_animation_frame(callback)
         .map_err(|err| anyhow!("Cannot request animation
                              frame {:#?}", err))
}
```

The `Function` type is a pure JavaScript type and is only available in the `js-sys` package. While we could import that, I'd rather not add another crate dependency if possible; however, we don't actually have to use the `Function` type directly if we make a small change to the function signature and the implementation:

```
pub fn request_animation_frame(callback: &Closure<
  dyn FnMut(f64)>) -> Result<i32> {
    window()?
         .request_animation_frame(callback.as_ref().unchecked_
ref())
         .map_err(|err| anyhow!("Cannot request animation
                              frame {:#?}", err))
}
```

Instead of taking `&Function`, our `request_animation_frame` will take `&Closure<dyn FnMut(f64)>` as its parameter. Then, it will call `callback. as_ref().unchecked_ref()` when calling the `web-sys` version of `request_animation_frame`. This converts `Closure` into `Function`, without requiring an explicit dependency on the `Function` type, and it's worth thinking about when you're creating your own versions of these functions. The makers of `web-sys` have to match every single potential use case, and as such, they are going to create the widest possible interfaces. As an application programmer, you do not need most of what's in that library. Therefore, you can and should narrow the interface to your own use cases, making it easier for you to work with. In fact, in order to make things a little cleaner, we'll convert that into a type, with one small change:

```
pub type LoopClosure = Closure<dyn FnMut(f64)>;
pub fn request_animation_frame(callback: &LoopClosure) ->
```

```
    Result<i32> {
// ...
```

Moving on from my mini-rant, you might think you can now write a simple game loop, like so:

```
pub fn animate(perf: f64) {
    browser::request_animation_frame(animate);
}
```

Ah, if only, but remember that we need to pass a JavaScript `Closure`, not a Rust `fn`. Using the `Closure::once` that we used before won't work because this closure will be called more than once, but fortunately, there's `Closure::wrap`, which will do just that. We'll create a function in `browser` to create a `Closure` specific to the `request_animation_frame` function, called `create_raf_closure`:

```
pub fn create_raf_closure(f: impl FnMut(f64) + 'static) ->
    LoopClosure {
    closure_wrap(Box::new(f))
}
```

The function being passed in has a `'static` lifetime. Any `FnMut` passed into this function cannot have any non-static references. This wasn't a decision on my part; it's a requirement of the `Closure::wrap` function we'll be calling into.

> **Tip**
> For more information on static lifetimes, take a look at the *Rust by Example* book, available for free here: `https://doc.rust-lang.org/rust-by-example/scope/lifetime/static_lifetime.html`.

Speaking of `Closure::wrap`, let's wrap it in a `closure_wrap` function so that the code we just added will compile, which looks like the following:

```
pub fn closure_wrap<T: WasmClosure + ?Sized>(data: Box<T>)
    -> Closure<T> {
    Closure::wrap(data)
}
```

This is another one of those wrapper functions where we are just matching the same signature as the function being wrapped – `Closure::wrap`. Because the `wrap` function on `Closure` creates a `Closure` that can be called multiple times, it needs to be wrapped in a `Box` and stored on the heap.

> **Tip**
>
> The nightly build of `wasm-bindgen` provides a more ergonomic `new` function that handles the boxing for you. We'll stick to the stable build in this book, but you are welcome to try nightly.

Now that you know the basic game loop and how to call `request_animation_frame`, you might think, "I've got this" and create the game loop as follows:

```
let animate = create_raf_closure(move |perf| {
    request_animation_frame(animate);
});

request_animation_frame(animate);
```

This is closer, but it's not there yet. Remember earlier that the `Closure` we pass to `create_raf_closure` has to have a `'static` lifetime, meaning everything that the `Closure` references must be owned by the closure. That's not the case right now. The `animate` variable is owned by the current scope and will be destroyed when that scope completes. Of course, `animate` is itself the `Closure` because this is a self-referencing data structure. The `animate` variable is the `Closure` but is also referenced inside the `Closure`. This is a classic Rust problem because the `borrow` checker cannot allow it.

Imagine what would happen if this wasn't the case – if `animate` could be referenced in the `Closure` but be owned by the scope outside the `Closure`. It would be destroyed when the program exited this scope, and `Closure` would no longer be valid – a `Null` pointer error, and a crash. This is the trouble with a self-referencing data structure, so we'll need a way to work around the `borrow` checker.

With nowhere to put this code just yet, let's have another crack at a hypothetical loop:

```
let f = Rc<RefCell<Option<LoopClosure>>> =
    Rc::new(RefCell::new(None));
let g = f.clone();

let animate = Some(create_raf_closure(move |perf: f64| {
```

```
    request_animation_frame(f.borrow().as_ref().unwrap());
});
```

```
*g.borrow_mut() = animate;
request_animation_frame(g.borrow().as_ref().unwrap());
```

Right now, I kinda wish I was writing JavaScript, but let's work slowly through this code. What we're doing is creating two references to the same place in memory, using Rc struct, allowing us to both take f and g and point them at the same thing but also move f into animate Closure. The other trick is that they both point to Option so that we can move f into Closure before it is completely defined. Finally, when we assign to g the Closure with *g.borrow_mut() = animate, we **also** assign to f because they are pointing to the same place. Did you get all that? No, me neither. Let's go through the types really quickly to reiterate what we did. f is set to the following:

- Rc to create a reference-counted pointer

- RefCell to allow for interior mutability

- Option to allow us to assign f to None

- LoopClosure to hold a mutable Closure that matches the request_ animation_frame parameter

g is then set to a clone of f so that they point to the same thing, and f is moved into animate Closure. g is assigned to animate via the dereference * operator and borrow_mut functions. Because f points to the same place as g, it will also contain animate Closure. Finally, we can call request_animation_frame, both outside and inside Closure, by borrowing it, converting it to a reference, and calling unwrap to actually get the real Closure. Yes, unwrap is back; we'll deal with one of those when we create our real function. Finally, g can be destroyed when it leaves scope because f is still in Closure and will keep the memory around.

> **Important Note**
> Once again, I'd love to take credit for this code, but the truth is that it's largely defined in the wasm-bindgen guide at https://bit.ly/3v5FG3j.

Now that we know what the core of our game loop is going to look like, how do we integrate it with a game?

A game trait

To write our game loop, we have a few options. We could just write the game in the loop, but that would look suspiciously similar to what we started with before. We could create a GameLoop struct with functions for update and draw, which is a significant improvement but still ties everything into one structure. We're going to go slightly beyond that and take inspiration from a popular game framework, XNA, or MonoGame in its modern incarnation. In the XNA framework, the game developer will implement a Game type, with methods for update and draw. This is slightly more complex than jamming all the code into one place but is significantly less so than a complete entity-component framework. It should work well for our purposes since it starts small and should allow for expansion as the game gets larger. There's a reason XNA was very successful.

> **Important Note**
>
> You can learn about XNA's modern equivalent, MonoGame, at https://www.monogame.net/.

We'll create a start function that accepts anything that implements the Game trait. The Game trait will start with two functions, update and draw. We'll run that through our game loop to first update and then draw our scene. All of this will go into the engine module; indeed, arguably, this is our entire "engine." Let's start with the simple version – first, the trait:

```
pub trait Game {
    fn update(&mut self);
    fn draw(&self, context: &CanvasRenderingContext2d);
}
```

So far so good. Note how the draw function takes CanvasRenderingContext2d as a parameter. Now for the rest of the loop – you can add this after the Game trait or load_image; it doesn't really matter as long as it's in the engine module:

```
pub struct GameLoop;
type SharedLoopClosure = Rc<RefCell<Option<LoopClosure>>>;

impl GameLoop {
    pub async fn start(mut game: impl Game + 'static) ->
        Result<()> {
            let f: SharedLoopClosure =
                Rc::new(RefCell::new(None));
```

```
        let g = f.clone();

    *g.borrow_mut() = Some(
      browser::create_raf_closure(move |perf: f64| {
        game.update();
        game.draw(&browser::context().expect("Context
          should exist"));

        browser::request_animation_frame(
          f.borrow().as_ref().unwrap());
    }));

    browser::request_animation_frame(
        g.borrow()
            .as_ref()
            .ok_or_else(|| anyhow!("GameLoop: Loop is
                                    None"))?,
    )?;
    Ok(())
  }
}
```

This is a bit larger but it's nothing you haven't seen before. We're going to create a GameLoop struct with no data and add a SharedLoopClosure type to simplify the type of the f and g variables. Then, we'll add an implementation of GameLoop with one method, start, that takes the Game trait as a parameter. Note that the trait is 'static because anything moved into the "raf" closure has to be 'static. We follow the snippets we used before to set up our request_animation_frame loop, and the key change is on the inside where we update and then draw, passing the draw function CanvasRenderingContext2d.

There's a problem with this kind of naive game loop. Typically, request_animation_frame runs at 60 frames per second, but if either update or draw takes longer than 1/60th of a second, it will slow down, making the game move more slowly. A long time ago, I recall beating levels by turning off the "**Turbo**" button on my desktop, making it possible to beat previously impossible challenges because the game became easier to play at slower speeds. Since we want a consistent experience across processor speeds, we'll take a common approach called "fixing" the time step.

Fixing our time step

You might notice that the update function we wrote doesn't take perf as a parameter; in fact, it's unused. Now, imagine trying to simulate a dog running across the screen, with no knowledge of how much time has passed between frames. Depending on the computer and your guess, the dog could saunter from left to right, or shoot past like a bullet. What we could do is send the delta time on each update, which can work but gets complicated very quickly. Instead, we'll assume every single tick takes the same amount of time, 1/60th of a second, and call update several times to "catch up" if we fall behind. It looks like this:

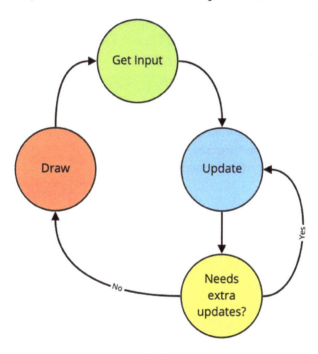

Figure 3.4 – A fixed step game loop

This isn't a perfect solution; if our game is very slow, it'll grind to a halt, but it should be good enough for our purposes. This is why I had us create a GameLoop struct – to track the time of the last update. We'll add two fields to the GameLoop struct:

```
const FRAME_SIZE: f32 = 1.0 / 60.0 * 1000.0;


pub struct GameLoop {
    last_frame: f64,
    accumulated_delta: f32,
}
```

This adds a constant for the length of a frame, converted to milliseconds. We'll track when the previous frame was requested in the `last_frame` field, and we'll accumulate a delta that totals up the physics time since the last render. It's not quite the same thing twice, as you'll see when we implement that counter in the `start` function. Speaking of that function, we'll need to initialize a mutable `GameLoop` at the beginning of that function:

```
impl GameLoop {
    pub async fn start(mut game: impl Game + 'static) ->
    Result<()> {
        let mut game_loop = GameLoop {
            last_frame: browser::now()?,
            accumulated_delta: 0.0,
        };
        ...
```

This initializes `GameLoop` appropriately, using now as the time of the last frame instead of 0 so that our loop doesn't perform several million updates before the first render. `browser::now()` hasn't been implemented yet, so you'll need to add it to the `browser` module:

```
pub fn now() -> Result<f64> {
    Ok(window()?
        .performance()
        .ok_or_else(|| anyhow!
            ("Performance object not found"))?
        .now())
}
```

This is just a wrapper around the web browser's now function. If you've been diligently following, you will probably recognize that this causes a compiler error. You'll need to add the "Performance" feature flag to the web-sys features list to bring in that function.

Now that we've created a game loop object, inside the `request_animation_frame` closure, we'll add our accumulator:

```
*g.borrow_mut() = Some(browser::create_raf_closure(move
    |perf: f64| {
      game_loop.accumulated_delta += (
          perf - game_loop.last_frame) as f32;
      while game_loop.accumulated_delta > FRAME_SIZE {
```

```
        game.update();
        game_loop.accumulated_delta -= FRAME_SIZE;
    }
    game_loop.last_frame = perf;
    game.draw(&browser::context().expect("Context should
      exist"));
```

What's changed since last time is that instead of just calling the `update` function immediately, we calculate the difference between `perf`, which if you remember from earlier is a high-res timestamp of the time that the `request_animation_frame` function started executing callback functions. We get the difference between now (in `perf`) and the previous frame and add that to `accumulated_delta`. Then, we compare this to our desired `FRAME_SIZE` (that's 1/60th of a second), and if there's **more** accumulated delta than the frame size, we call `update`. Then we subtract the frame size from the delta. What is the effect of all this? If `game.draw` takes too long so that we cannot complete 1 frame in 1/60th of a second, the code will run extra updates to catch up.

An example is helpful here. Assume you started playing the game at time 0, the beginning of the world. When the first callback executes for `request_animation_frame` its probably very close to 0, perhaps as low as 1 millisecond, because there's no delay on the first frame. The code will add that to `accumulated_delta` and then compare it to `FRAME_SIZE` and see that there hasn't been enough delta accumulation, so `update` is skipped. The `last_frame` value is stored (again, we'll say it's 1), the screen is drawn, and then `request_animation_frame` is called.

The second time though, the value of `perf` is likely to be about the size of the first frame. We'll use 17 milliseconds for simple math. So `perf` is 17; subtract from it the `last_frame`, which is 1, and add 16 milliseconds to `accumulated_delta`. The new value of `accumulated_delta` is 17, so the game is updated once and `accumulated_delta` is reduced to 1. The game continues with one update to one draw until something goes wrong. The `draw` call takes 40 milliseconds! Who knows why – maybe an autoplay video started up by surprise, taking resources. It doesn't matter because `accumulated_delta` shoots up to 40, which is larger than 2 frames. Now, the loop on `accumulated_delta` runs `update` twice, dropping a frame of animation to compensate for the drop in performance. The important thing to remember here is that it drops a *draw* but not an *update*, so while the player might see some visual artifacts, the physics will still work without issue.

> **Important Note**
>
> You might wonder what happens to the extra `accumulated_delta` since it's unlikely to be an exact multiple of `FRAME_SIZE`. More advanced game loops will pass that into the draw and use it to interpolate between the two update values. We shouldn't need that for our game and will just roll that `delta` over to the next frame.

> **Important Note**
>
> Why use an `f32` for the `accumulated_delta`? That's very observant of you! The short version is, because we can. The slightly longer version is that we only use `f64` as often as we do because JavaScript uses a 64-bit `Number` type for all its numbers. If I could, I'd use smaller values whenever possible, and integers as well, because the extra size of `f64` isn't really necessary and can cause a surprising drag on performance when repeated everywhere.

So, there you have it – your game loop, at least the "looping" part of it. While it's usable now, it doesn't provide an easy way to load our assets. While we could just leave things as they are and always load assets before we start our game loop, a cleaner solution is to integrate that rule into the game loop.

Loading assets

Expanding our game loop to handle loading assets is going to require adding a function to our trait, an `async` one to be precise. This will allow us to put all our asynchronous code that's currently wrapped in the `spawn_local` in `lib` and put it in a function that returns `Result` with `Game` in it. You can start by adding that function to the `Game` trait:

```
pub trait Game {
    async fn initialize(&self) -> Result<Box<dyn Game>>;
    fn update(&mut self);
    fn draw(&self, context: &Renderer);
}
```

Unfortunately, that doesn't compile. `async` trait functions haven't landed in stable Rust yet, but fortunately, we can use a crate to get that functionality. Add `async-trait = "0.1.52"` to `Cargo.toml` and then add the following attribute macro to the trait:

```
#[async_trait(?Send)]
pub trait Game {
```

You'll also need to import `async_trait::async_trait`. The `async_trait` allows us to add `async` functions to a trait. We can use it with the `?Send` trait because we don't need our futures to be thread-safe. Now, we can add this to the game loop:

```
impl GameLoop {
    pub async fn start(game: impl Game + 'static) ->
      Result<()> {
        let mut game = game.initialize().await?;
        ....
```

And that's it! The game gets initialized, asynchronously, and with `Result` on the first line. Note that the passed-in `game` no longer has to be mutable since we're not mutating it in the function anywhere. We're almost ready to integrate our old `set_interval` into this, but there's a little more cleanup I'd like to do around drawing.

Cleaner drawing

Currently, we're sending a raw `CanvasRenderingContext2d` to the draw loop, with all of its awkward functions such as `draw_image_with_html_image_element_and_sw_and_sh_and_dx_and_dy_and_dw_and_dh`. This works but it's ugly, and much like we did with the `browser` module, we can use a wrapper to narrow the context's wide interface to a smaller one, tailored to our needs. We'll replace passing `CanvasRenderingContext2d` with our own `Renderer` object that has easier-to-use functions.

We'll start by creating a structure for our `Renderer` in `engine`:

```
pub struct Renderer {
    context: CanvasRenderingContext2d,
}
```

This is a simple wrapper containing the rendering context. For now, we'll just add the two implementation methods to the `Renderer` struct:

```
impl Renderer {
    pub fn clear(&self, rect: &Rect) {
        self.context.clear_rect(
            rect.x.into(),
            rect.y.into(),
            rect.width.into(),
            rect.height.into(),
```

```
        );
    }


    pub fn draw_image(&self, image: &HtmlImageElement,
        frame: &Rect, destination: &Rect) {
        self.context
            .draw_image_with_html_image_element_and_sw_and_sh_
                and_dx_and_dy_and_dw_and_dh(
                &image,
                frame.x.into(),
                frame.y.into(),
                frame.width.into(),
                frame.height.into(),
                destination.x.into(),
                destination.y.into(),
                destination.width.into(),
                destination.height.into(),
            )
            .expect("Drawing is throwing exceptions!
                Unrecoverable error.");
    }
}
```

These two functions, `clear` and `draw_image`, both wrap `context` functions but do so using fewer parameters. Instead of four parameters and `clear_rect`, we pass `clear` Rect. Instead of that incredibly long function name, we pass `draw_image` `HtmlImageElement` and two `Rect` structures. Currently, we go ahead and use `expect` to panic! here if we can't draw. I am convinced that this should return `Result`.

> **Important Note**
>
> By now, there's been code in this book that you've thought could be done better. Try that out! I do that all the time when I follow books, and there's no reason you shouldn't too. Just try to remember where you've diverged from the book.

Of course, both of these functions take Rect, but we don't have a Rect structure. Let's add that to the engine now:

```
pub struct Rect {
    pub x: f32,
    pub y: f32,
    pub width: f32,
    pub height: f32,
}
```

Now we can change the draw function to take Renderer instead of CanvasRenderingContext2d. So, we update the trait:

```
#[async_trait(?Send)]
pub trait Game {
    ...
    fn draw(&self, renderer: &Renderer);
}
```

Then, we can make a change to the loop. Right now, we create context in the Closure that we pass to create_raf_closure. That call returns Result, so to get access to context, we have to call unwrap or expect. The cleaner approach we can use now is to create Renderer, with Context outside of Closure, as shown here:

```
let mut game_loop = GameLoop {
    last_frame: browser::now()?,
    accumulated_delta: 0.0,
};

let renderer = Renderer {
    context: browser::context()?,
};
...
*g.borrow_mut() = Some(browser::create_raf_closure(
  move |perf: f64| {
    ...
    game.draw(&renderer);
```

```
browser::request_animation_frame(f.borrow().as_ref().
    unwrap());
}));
```

Moving this outside of the `request_animation_frame` closure means we don't need to use the `expect` syntax anymore – nice!

The small change to `draw`, turning it into `game.draw(&renderer)`, will make our `draw` function easier to write. I think we're meeting our goal of changing the code to make it *easier* to move forward. Let's prove it by taking our animation code out of `lib` and using the game loop.

Integrating the game loop

It's great that we've written this game loop and all, but it's about time we actually use it. Remember that we have our `GameLoop` structure, but it operates on a `Game` trait. So in order to use the loop, we need to implement that trait. We'll implement it in another module, game, which we'll create in `game.rs` and then add to the library using the mod game instruction declaration in `lib.rs`. We'll start with a couple of structures:

```
use crate::engine::{Game, Renderer};
use anyhow::Result;
use async_trait::async_trait;


pub struct WalkTheDog;


#[async_trait(?Send)]
impl Game for WalkTheDog {
    async fn initialize(&self) -> Result<Box<dyn Game>> {
        Ok(Box::new(WalkTheDog {}))
    }
    fn update(&mut self) {}
    fn draw(&self, renderer: &Renderer) {}
}
```

Make sure that you add the `#[async_trait(?Send)]` annotation, which allows you to implement a trait with the `async` functions. Provided you add the required `use` declarations from `engine`, this compiles because `Game` implements the trait as needed. It doesn't do anything, but it compiles. The `initialize` function might look a little strange because we're taking `self` and just throwing it away in favor of a new `WalkTheDog` structure – thrown on the heap, no less! We're doing that for some changes that you'll see in the next chapter, so just bear with me for now.

Now, let's take the code that draws from `lib.rs` and move it into `draw`, updating it along the way:

```
fn draw(&self, renderer: &Renderer) {
    let frame_name = format!("Run ({}).png", self.frame +
                             1);
    let sprite = self.sheet.frames.get(&frame_name).expect(
        "Cell not found");

    renderer.clear(Rect {
        x: 0.0,
        y: 0.0,
        width: 600.0,
        height: 600.0,
    });
    renderer.draw_image(
        &self.image,
        Rect {
            x: sprite.frame.x.into(),
            y: sprite.frame.y.into(),
            width: sprite.frame.w.into(),
            height: sprite.frame.h.into(),
        },
        Rect {
            x: 300.0,
            y: 300.0,
            width: sprite.frame.w.into(),
            height: sprite.frame.h.into(),
```

```
        },
    );
}
```

This only contains slight changes to the code in lib.rs, although it definitely won't compile. Calls to context are replaced with calls to renderer, and we've used the new Rect structure. This won't compile because self doesn't have sheet, frame, or image. We'll need to add that to the game module, as follows:

```
#[derive(Deserialize)]
struct SheetRect {
    x: i16,
    y: i16,
    w: i16,
    h: i16,
}

#[derive(Deserialize)]
struct Cell {
    frame: SheetRect,
}

#[derive(Deserialize)]
pub struct Sheet {
    frames: HashMap<String, Cell>,
}

pub struct WalkTheDog {
    image: HtmlImageElement,
    sheet: Sheet,
    frame: u8,
}
```

Here, we've moved the structures from `lib.rs` that serialize the JSON from our sprite sheet and added fields for `frame`, `HtmlImageElement`, and `Sheet` to the `WalkTheDog` struct. Pay close attention to the fact that we've taken `Rect` from `lib` and renamed it `SheetRect`. This is the specific rectangle from our sprite sheet. In `game`, we also have a `Rect` structure. This is the rectangle that we'll use as a game domain object. This rename is confusing right now but is done to differentiate the two rectangles and is helpful as we go forward.

The `WalkTheDog` structure has the fields needed to make `draw` compile, but it may make you wonder about `initialize`. Specifically, if we're going to move our loading code to `initialize`, does the `WalkTheDog` struct really always have `HtmlImageElement` and `Sheet`? No, it does not. We'll need to convert those fields to `Option` types and make the `draw` function account for them:

```
pub struct WalkTheDog {
    image: Option<HtmlImageElement>,
    sheet: Option<Sheet>,
    frame: u8,
}
```

We can use the `as_ref()` function to borrow `image` and `sheet`, and then use the `and_then` and `map` `Option` functions to cleanly get the frame and then draw it:

```
fn draw(&self, renderer: &Renderer) {
    let frame_name = format!("Run ({}).png", self.frame + 1);
    let sprite = self
        .sheet
        .as_ref()
        .and_then(|sheet| sheet.frames.get(&frame_name))
        .expect("Cell not found");
    renderer.clear(&Rect {
        x: 0.0,
        y: 0.0,
        width: 600.0,
        height: 600.0,
    });

    self.image.as_ref().map(|image| {
        renderer.draw_image(
```

```
            &image,
            &Rect {
                x: sprite.frame.x.into(),
                y: sprite.frame.y.into(),
                width: sprite.frame.w.into(),
                height: sprite.frame.h.into(),
            },
            &Rect {
                x: 300.0,
                y: 300.0,
                width: sprite.frame.w.into(),
                height: sprite.frame.h.into(),
            },
        );
    });
```

This is great – we've got a game that draws absolutely nothing, but that's okay since our initialize code still doesn't compile. Let's prepare to draw by copying our loading code from lib.rs to the initialize function in the game loop. Don't do any cutting and pasting yet; we'll go ahead and clean up lib.rs at the end. Initialize should now look like this:

```
impl Game for WalkTheDog {
    async fn initialize(&self) -> Result<Box<dyn Game>> {
        let sheet: Sheet = browser::fetch_json("rhb.json")
            .await
            .expect("Could not fetch rhb.json")
            .into_serde()
            .expect("Could not convert rhb.json into a
                Sheet structure");

        let image = engine::load_image("rhb.png")
            .await
            .expect("Could not load rhb.png");

        Ok(Box::new(WalkTheDog {
            image: Some(image),
```

```
            sheet: Some(sheet),
            frame: self.frame,
        }))
    }
    . . .
```

That's a great copy and paste, but we can make it far more concise by using the ? operator. Here's the version with that improvement:

```
async fn initialize(&self) -> Result<Box<dyn Game>> {
    let sheet = browser::fetch_json(
        "rhb.json").await?.into_serde()?;

    let image =
        Some(engine::load_image("rhb.png").await?);

    Ok(Box::new(WalkTheDog {
        image,
        sheet,
        frame: self.frame,
    }))
}
```

Look how small and clean that function is. It only took us three tries, but we got there. Now that we have initialize and draw, we can write update. The version we wrote in lib.rs used set_interval_with_callback_and_timeout_and_ arguments_0 to animate our Red Hat Boy, but that's not going to work anymore. Instead, the update function will need to keep track of the number of frames that have passed and advance when it's appropriate. In the original code, we called the set_ interval callback every 50 milliseconds. In this new code, update will be called every 1/60th of a second, or 16.7 milliseconds. So, in order to approximately match the animation, we'll want to update the current sprite frame every three updates; otherwise, our little RHB will run very, very fast.

If you look at the rhb.json file, you can see that there are eight frames in the Run animation. If we want to advance a sprite frame every 3 updates, that means it will take 24 updates to complete the animation. At that point, we'll want to return to the beginning and play it again. So, we'll need to calculate the sprite frame from the frame count, which is updated in the update function:

```
fn update(&mut self) {
    if self.frame < 23 {
        self.frame += 1;
    } else {
        self.frame = 0;
    }
}
```

This won't work with our current draw code because it uses frame to look up the sprite to render. It will crash when it looks for Run (9).png, which doesn't exist. We'll update the draw function to get the sprite index from frame:

```
fn draw(&self, renderer: &Renderer) {
    let current_sprite = (self.frame / 3) + 1;
    let frame_name = format!("Run ({}).png",
      current_sprite);
    ...
```

The current_sprite variable will cycle from one to eight, and then loop back again. Don't believe me? Feel free to use the log! macro we wrote earlier to check my work; in fact, I encourage you to. Not because I'm arrogant but because it's always good to experiment with the code, rather than blindly typing it in. We then take that number and use it to look up the frame name.

With that accomplished, we now have a game loop that can render to the canvas and a game that renders our running RHB; we just need to integrate it. We'll add a plain constructor to the WalkTheDog struct, right under the struct definition in engine:

```
impl WalkTheDog {
    pub fn new() -> Self {
        WalkTheDog {
            image: None,
            sheet: None,
            frame: 0,
```

```
            }
        }
    }
```

Nothing spectacular there – just something to make it easier to create the game object. And now for the moment you've been waiting for – the new main function integrating all these changes:

```
#[wasm_bindgen(start)]
pub fn main_js() -> Result<(), JsValue> {
    console_error_panic_hook::set_once();

    browser::spawn_local(async move {
        let game = WalkTheDog::new();

        GameLoop::start(game)
            .await
            .expect("Could not start game loop");
    });

    Ok(())
}
```

No, really, that's it – that's the whole thing. You spawn a local future, create a new game, and then call GameLoop::start(game).await to start it up. You can delete all the unused code from lib.rs, such as the extra use declarations and the structures we defined when everything was here. It looks great!

We changed a lot of code to get here, but now we have a running game with a proper loop. We could end the chapter here, but it would be kind of nice if the code actually did something new, wouldn't it?

Adding keyboard input

Most games have some form of user input; otherwise, they aren't much of a game. In this section, we'll start listening to keyboard events and use them to control our RHB. That means adding keyboard input to the game loop and passing that into the update function. What we will *not* be doing is yet more refactoring. The system is reasonably well factored at this point and is open to our new changes.

The specific process by which we'll get keyboard events is probably a little different than you're used to if you do web development. In a normal program, you would listen for keys to get pressed – in other words, pushed down and then released – and then do something such as update the screen when the button is released. This doesn't fit in with a game because typical players want the action to happen as soon as a key is pushed down and want it to continue for as long as it's held. Think of moving around the screen with the arrow keys. You expect motion to start the second you hit the arrow key, not after you release it. In addition, traditional programming doesn't account for things like pressing "up" and "right" at the same time. If we process those as two separate actions, we'll move right, then up, then right, and then up, like we're moving up the stairs. What we'll do is listen to every keyup and keydown event, and bundle that all up into a keystate that stores every currently pressed key. Then we'll pass that state to the update function so that the game can figure out just what to do with all the currently pressed keys.

> **Important Note**
>
> This approach is common in games, and it leads to one downside. If you want to trigger something only when a button is pressed, such as firing a gun, you have to keep track of whether or not the previous update had the key up and the next update had it down. So, by flipping from an event-driven approach to a global key state, we lose the events. Fortunately, this is easily recreated.

To get keyboard events, we have to listen for the keydown and keyup events on canvas. Let's start with a new function in engine, prepare_input():

```
fn prepare_input() {
    let onkeydown = browser::closure_wrap(
        Box::new(move |keycode: web_sys::KeyboardEvent| {})
            as Box<dyn FnMut(web_sys::KeyboardEvent)>);

    let onkeyup = browser::closure_wrap(Box::new(
        move |keycode: web_sys::KeyboardEvent| {})
```

```
            as Box<dyn FnMut(web_sys::KeyboardEvent)>);

    browser::canvas()
        .unwrap()
        .set_onkeydown(Some(onkeydown.as_ref().unchecked_
ref()));
    browser::canvas()
        .unwrap()
        .set_onkeyup(Some(onkeyup.as_ref().unchecked_ref()));
    onkeydown.forget();
    onkeyup.forget();
}
```

> **Tip**
>
> Make sure you set up your `canvas` element with a `tabIndex` attribute in the HTML file; otherwise, it cannot get focus and have keyboard events.

This is enough to get us started. It should look familiar because we're setting up `Closure` objects in the same way we did for `load_image` and `request_animation_frame`. We have to make sure we call `forget` on both of the `Closure` instances so that they aren't deallocated immediately after being set up because nothing in the Rust application is holding onto them. You'll also need to add the `KeyboardEvent` feature to `web-sys` to include it. Otherwise, there is nothing here you haven't seen before. It just doesn't do anything yet.

> **Tip**
>
> Unlike most things in Rust, if you don't add a `forget` call, you won't get a compile-time error. You'll get a panic almost immediately and not always with a helpful error message. If you think you've set up callbacks into JavaScript and you're getting panics, ask yourself whether anything is holding on to that callback in your program. If nothing is, you've probably forgotten to add `forget`.

We're listening to the input, so now we need to keep track of all of it. It's tempting to start trying to condense the events into keystate in this function, but that's troublesome because this function only handles one keyup or keydown at a time and doesn't know anything about all the other keys. If you wanted to keep track of an ArrowUp and ArrowRight being pressed at the same time, you couldn't do it here. What we will do is set up the listeners once before the game loop starts, such as with initialize, and then process all the new key events on every update updating our keystate. This will mean sharing state from these closures with the closure we passed to request_animation_frame. It's time to add a channel. We'll create an unbounded channel, which is a channel that will grow forever if you let it, here in prepare_input and then return its receiver. We'll pass transmitters to both onkeyup and onkeydown, and send the KeyboardEvent to each of those. Let's take a look at the changes:

```
fn prepare_input() -> Result<UnboundedReceiver<KeyPress>> {
    let (keydown_sender, keyevent_receiver) = unbounded();
    let keydown_sender = Rc::new(RefCell::new(keydown_sender));
    let keyup_sender = Rc::clone(&keydown_sender);
    let onkeydown = browser::closure_wrap(Box::new(move
|keycode: web_sys::KeyboardEvent| {
        keydown_sender
            .borrow_mut()
            .start_send(KeyPress::KeyDown(keycode));
    }) as Box<dyn FnMut(web_sys::KeyboardEvent)>);

    let onkeyup = browser::closure_wrap(Box::new(move |keycode:
web_sys::KeyboardEvent| {
        keyup_sender
            .borrow_mut()
            .start_send(KeyPress::KeyUp(keycode));
    }) as Box<dyn FnMut(web_sys::KeyboardEvent)>);

    browser::window()?.set_onkeydown(Some(onkeydown.as_ref().
unchecked_ref()));
    browser::window()?.set_onkeyup(Some(onkeyup.as_ref().
unchecked_ref()));
    onkeydown.forget();
    onkeyup.forget();
```

```
        Ok(keyevent_receiver)
}
```

The function now returns `Result<UnboundedReceiver<KeyPress>>`.
`UnboundedReceiver` and `unbounded` are both in the `futures::channel::mspc`
module and are declared in a `use` declaration at the top of the file. We create the
unbounded channel on the first line with the `unbounded` function and then create
reference counted versions of both `keydown_sender` and `keyup_sender`, so that we
can move each of them into their respective closures while sending both events to the
same receiver. Note that the unbounded channel uses `start_send` instead of `send`.
Finally, we return `keyevent_receiver` as `Result`. You might consider having two
independent channels, one for `keyup` and one for `keydown`, and while I'm certain that
can be done, I tried it and found this way was more straightforward.

Look closely and you might wonder what `KeyPress` is. It turns out you can't tell what
kind of `KeyboardEvent` happened simply by inspecting it. In order to keep track of
whether the event was `keyup` or `keydown`, we wrap those events in an enumerated type
that we'll define in `engine.rs`:

```
enum KeyPress {
    KeyUp(web_sys::KeyboardEvent),
    KeyDown(web_sys::KeyboardEvent),
}
```

This enum approach means we won't have to manage two channels. Now that we have
a function that will listen for and put all our key events into a channel, we need to write
a second function that grabs all those events off the channel and reduces them into
`KeyState`. We can do that like so, still in the `engine` module:

```
fn process_input(state: &mut KeyState, keyevent_receiver: &mut
UnboundedReceiver<KeyPress>) {
    loop {
        match keyevent_receiver.try_next() {
            Ok(None) => break,
            Err(_err) => break,
            Ok(Some(evt)) => match evt {
                KeyPress::KeyUp(evt) => state.set_
released(&evt.code()),
                KeyPress::KeyDown(evt) => state.set_
pressed(&evt.code(), evt),
            },
```

```
        };
    }
}
```

This function takes `KeyState` and `Receiver` and updates `state` by taking every entry off of the receiver until its empty. Theoretically, this appears to create the possibility for an infinite loop in the event that the receiver is constantly filled, but I was unable to do that by normal means (pressing the keyboard like a madman), and if somebody decides to write a script that fills this channel and break their own game, more power to them. `KeyState` has to be passed as `mut` so that we update the current one and do not start from a brand-new state on each update. We've written this function pretending that `KeyState` already exists, but we need to create it as well, again in the `engine` module:

```
pub struct KeyState {
    pressed_keys: HashMap<String, web_sys::KeyboardEvent>,
}

impl KeyState {
    fn new() -> Self {
        KeyState {
            pressed_keys: HashMap::new(),
        }
    }

    pub fn is_pressed(&self, code: &str) -> bool {
        self.pressed_keys.contains_key(code)
    }

    fn set_pressed(&mut self, code: &str, event: web_
sys::KeyboardEvent) {
        self.pressed_keys.insert(code.into(), event);
    }

    fn set_released(&mut self, code: &str) {
        self.pressed_keys.remove(code.into());
    }
}
```

The `KeyState` struct is just a wrapper around `HashMap`, storing a lookup of `KeyboardEvent.code` to its `KeyboardEvent`. If the `code` isn't present, then the key isn't pressed. The code is the actual representation of a physical key on the keyboard. You can find a list of all the available `KeyboardEvent` codes on MDN Web Docs: `https://mzl.la/3ar9krK`.

> **Tip**
>
> When in doubt, MDN Web Docs from Mozilla is easily the best resource on the web for browser libraries.

We've created the libraries and structures we need for keyboard input, so now we can integrate it into our `GameLoop`. We'll call `prepare_input` in the `start` function before we start looping:

```
pub async fn start(mut game: impl Game + 'static) -> Result<()>
{
    let mut keyevent_receiver = prepare_input()?;
    game.initialize().await?;
```

Then, we'll move `keyevent_receiver` into the `request_animation_frame` closure and process the input on every update:

```
let mut keystate = KeyState::new();
*g.borrow_mut() = Some(browser::create_raf_closure(move |perf:
f64| {
    process_input(&mut keystate, &mut keyevent_receiver);
```

You can see that we initialized an empty `KeyState` right before the `request_animation_frame` closure,so that we can start with an empty one. Each frame will now call our `process_input` function and generate a new `KeyState`. That's all the changes we have to do to our game loop to keep track of `KeyState`. The only thing that's remaining is to pass it to our `Game` object so that it can be used. Some game implementations will store this as a global, but we'll just pass it to the `Game` trait. We'll update the trait's `update` function to accept `KeyState`:

```
pub trait Game {
    ...
    fn update(&mut self, keystate: &KeyState);
    ...
```

Now, we can pass `KeyState` to the `update` function on every loop:

```
while game_loop.accumulated_delta > frame_size {
    game.update(&keystate);
    game_loop.accumulated_delta -= frame_size;
}
```

Finally, to keep our game compiling, we will need to update the `WalkTheDog::update` signature, over in the `game` module, to match:

```
#[async_trait(?Send)]
impl Game for WalkTheDog {

    ...

    fn update(&mut self, keystate: &KeyState) {
```

That's it! We've got a `GameLoop` that processes keyboard input and passes that state to our `Game`. We've spent a lot of time writing code that makes it possible for us to write a game, but we haven't actually updated our game. Our poor little RHB still just runs in one place. He looks happy, but now that we've got input, how about we move him around?

Moving Red Hat Boy

Moving game objects means keeping track of a position instead of hardcoding it, as you might have expected. We'll create a `Point` structure in `engine` that will hold an *x* and a *y* position for RHB. On every `update` call, we'll also calculate a velocity for him, based on which keys are pressed. Every direction will be the same size, so if `ArrowLeft` and `ArrowRight` are pressed at the same time, he'll stop moving. After we calculate his velocity, we'll update his position with that number. That should be enough to allow us to move him around the screen. Let's start by adding `position` to the `WalkTheDog` game struct:

```
pub struct WalkTheDog {
    image: Option<HtmlImageElement>,
    sheet: Option<Sheet>,
    frame: u8,
    position: Point,
}
```

Of course, `Point` doesn't exist yet, so we'll create it in `engine`:

```
#[derive(Clone, Copy)]
pub struct Point {
    pub x: i16,
    pub y: i16,
}
```

Note that we're using integers here so that we don't have to deal with floating point math when it's not necessary. While the `canvas` functions all take `f64` values, that's only because there is only one number type in `JavaScript`, and per MDN Web Docs (https://mzl.la/32PpIhL), canvas is faster if you use integer coordinates. You'll also need to update the `WalkTheDog::new` function to fill in a default `position`. Let's use `0, 0` for now:

```
impl WalkTheDog {
    pub fn new() -> Self {
        WalkTheDog {
            image: None,
            sheet: None,
            frame: 0,
            position: Point { x: 0, y: 0 },
        }
    }
}
```

I promised I would stop reminding you to do this, but do make sure you've added a `use` declaration for `crate::engine::Point` at the top of the file. The `initialize` function also needs to be updated to account for `position`. This is actually why we marked `Point` with `Clone` and `Copy`. It makes it possible to copy it into the new `WalkTheDog` initialize function, as shown here:

```
impl Game for WalkTheDog {
    async fn initialize(&self) -> Result<Box<dyn Game>> {
        let json = browser::fetch_json("rhb.json").await?;
        let sheet = json.into_serde()?;
        let image =
            Some(engine::load_image("rhb.png").await?);
        Ok(Box::new(WalkTheDog {
```

```
                image,
                sheet,
                position: self.position,
                frame: self.frame,
        }))
}
....
```

In order for `position` to have any meaning, we'll need to update the `draw` function so that it's actually being used:

```
#[async_trait(?Send)]
impl Game for WalkTheDog {
    ...
    fn draw(&self, renderer: &Renderer) {
        ....
        self.image.as_ref().map(|image| {
            renderer.draw_image(
                &image,
                &Rect {
                    x: sprite.frame.x.into(),
                    y: sprite.frame.y.into(),
                    width: sprite.frame.w.into(),
                    height: sprite.frame.h.into(),
                },
                &Rect {
                    x: self.position.into(),
                    y: self.position.into(),
                    width: sprite.frame.w.into(),
                    height: sprite.frame.h.into(),
                },
            );
        });
    }
}
```

Make sure you update the *second* Rect and not the first one. The first Rect is the slice we are taking out of our sprite sheet. The second one is where we want to draw it. This should cause a noticeable change to the game, as RHB is now in the upper-left corner. Finally, we're going to modify update to calculate a velocity based on which keys are pressed in KeyState. We'll add this before updating the current frame, as shown here:

```
fn update(&mut self, keystate: &KeyState) {
    let mut velocity = Point { x: 0, y: 0 };
    if keystate.is_pressed("ArrowDown") {
        velocity.y += 3;
    }

    if keystate.is_pressed("ArrowUp") {
        velocity.y -= 3;
    }

    if keystate.is_pressed("ArrowRight") {
        velocity.x += 3;
    }

    if keystate.is_pressed("ArrowLeft") {
        velocity.x -= 3;
    }
```

The "ArrowDown" and "ArrowUp" strings and so on are all listed at https://mzl. la/3ar9krK, although you can also figure them out by simply logging the code when a key is pressed. You can see here that if "ArrowDown" is pressed we increase y, and if "ArrowUp" is pressed, we decrease it, and that's because the origin is in the upper-left-hand corner, with y increasing as you go down, not up. Note also that we don't use if/ else here. We want to account for every pressed key and not short-circuit on the first key that's pressed. Next, we adjust the position based on velocity:

```
if keystate.is_pressed("ArrowLeft") {
    velocity.x -= 3;
}

self.position.x += velocity.x;
self.position.y += velocity.y;
```

Head back to the browser, and you can now use the arrow keys to move RHB around! If he doesn't move, make sure you click in the canvas to give it focus. If he still doesn't move and you're sure you've gotten everything right, put some `log!` messages in the `start` function and make sure `KeyState` is being created, or in the `update` function to see if you're actually getting a new `KeyState`. We've covered a lot of ground here, and if you're following along, it's very easy to make a mistake, but you have a debugging tool to figure out issues now.

> **Tip**
>
> On some browsers, the `canvas` will get a border around it when it has the focus, which will appear after you click it. You can remove that by adding a style of `outline: none` to the `canvas`.

Summary

This was a hard, long, and complicated chapter. I'll quote a phrase Aaron Hillegass uses frequently in his books: *"Programming is hard and you are not stupid."* There were plenty of areas where a small typo could trip you up, and you may have had to go backward and forward several times. That's all okay – it's part of the learning process. I would encourage you to experiment with the skeleton we've built, even before moving onto the next chapter, as it's a great way to ensure you understand all the code.

In the end, we've accomplished a lot. We've created a game loop that will run in the browser at 60 frames per second while updating at a fixed step. We've set up an XNA-like game "engine" and separated the engine concerns from the game concerns. Our browser interface is wrapped in a module so that we can hide some of the details of the browser implementation. We're even processing input, making this work like a true game engine. We did all this while keeping the code running as we went.

The code should be easier to work with going forward because we now have clear places to put things. Browser functions go in a browser, engine functions in an engine, and the game in a game module, although you might feel like it's not a game because RHB doesn't run, jump, and slide around.

Guess what we're doing next?

4
Managing Animations with State Machines

In the last chapter, we created a minimal game *engine*, allowing for moving our main character around and playing a simple animation, but it's far from full-featured. There's no world to navigate, the only animation that plays is running, and **Red Hat Boy** (**RHB**) doesn't respond to any physics. At this point, if we wanted to retitle our game, it would be called *Red Hat Boy and the Empty Void*.

While that might be a fun title, it wouldn't make for a fun game. Ultimately, we'll want RHB to chase his dog through a forest with platforms to jump on and slide under, and to do that we'll need to make sure he slides, jumps, and runs. We'll also need to make sure that he looks, acts, and behaves differently when he does those things.

In this chapter, we're going to introduce a common game development pattern to manage all that, the state machine, implemented in **Rust**. Rust gives us powerful constructs for state machines but also unique challenges due to its ownership model, so we'll dive into that and why we'll use it instead of deceptively simple `if` statements.

We're going to cover the following topics:

- Introducing state machines
- Managing animation
- Adding states to walk the dog
- `Idle`, `Running`, `Sliding`, and `Jumping` animations

By the end of the chapter, you will be able to use state machines to cleanly transition between animations while always playing the correct one.

Technical requirements

There are no new crates or other technical requirements in this chapter. The source code for this chapter is available at `https://github.com/PacktPublishing/Game-Development-with-Rust-and-WebAssembly/tree/chapter_4`.

Check out the following video to see the Code in Action: `https://bit.ly/35sk3TC`

Introducing state machines

Games, web applications, heck, even cryptocurrency miners, have to manage the *state* of the system. After all, if the system isn't doing something right now, if it doesn't have a current state, then it's not running, is it? The state is also fractal. In our game, we have a state of `playing`, and another one of `game over`. Once we add menu items, we'll have even more states. Meanwhile, our RHB also has states: he's running, sliding, jumping, dying, and dead. Let's say unconscious, that's less dark.

The point is our game is doing a lot of things and is maintaining a large game state with a lot of mini-states inside it. As the application moves from one state to another, the rules of the system change. For example, when RHB is running, the *spacebar* might make him jump, but when he's jumping, hitting the *spacebar* doesn't do anything. The rule is you can't jump when you're already jumping. One way you can maintain that state is through a large structure with a bunch of values or Booleans, such as `jumping = true`, and in a Rust program, you might store that in an enumerated type like this:

```
enum RedHatBoyState {
    Jumping,
    Running,
    Sliding,
}
```

This works reasonably well in small programs but for larger programs, there are two things that you'll want to manage. The first, which I've already hinted at, is that there may be rules about going between states. Maybe you can't go right from `Jumping` to `Sliding`, but an enum doesn't prevent that. The second is that, in addition to the rules being different for each state, frequently things happen on the *transitions* between states, things such as playing a sound effect or updating a score; for that, you need a state machine.

Defining a state machine

Perhaps the most confusing thing about state machines is the naming, as there are state machines, finite state machines, the state pattern, and more, all of which frequently get used interchangeably by programmers. So, for the sake of clarity, let's define them this way:

- **State machines**: A model of the state of a system, represented by a list of states and the transitions between them

- **State pattern**: One way to implement state machines, which we will *not* be using in our application, although our implementation will bear a resemblance to it

> **Important Note**
>
> The Rust Programming Language has an implementation of the traditional state pattern, using a `trait` object, which you can find here: `https://doc.rust-lang.org/book/ch17-03-oo-design-patterns.html`. It's quite good but is not idiomatic Rust, and we won't be using it.

The state machine both helps us keep a mental model of the system in our heads and prevents us from making foolish mistakes in code, such as playing the running animation while RHB is jumping. The drawback, of course, is that you need to understand state machines, so let's get that covered. We'll use RHB as our example. RHB can be **Running**, **Idle**, **Jumping**, **Sliding**, **Falling**, or **KnockedOut**. We can use a state **transition table** to list those:

State	Event	New State
Idle, Running		
Jumping		
Sliding		
Falling		
KnockedOut		

The transition table only has three columns for now, which are a start state, the event that causes a transition, and the state it transitions to. Events differ from transitions in that events are what happens to the system to *cause* a transition, but transitions are what happens *during* the state change.

It's a subtle difference, and sometimes it gets used interchangeably because the names will frequently be the same. Let's work through a state transition to clarify this. RHB starts in the Idle state, where he stands in place with an **Idle** animation. To start running, he gets an event; let's call it run:

State	Event	New State
Idle	run	Running
Running		
Jumping		
Sliding		
Falling		
KnockedOut		

When moving to Running, we actually do something on the transition. Specifically, we start moving to the right; we increase the velocity in x. You can name this transition in the table:

State	Event	New State	Transition
Idle	run	Running	IncreaseVelocity
Running			
Jumping			
Sliding			
Falling			
KnockedOut			

While this is correct, often we don't bother naming the transitions and the events because they become redundant. While we could continue adding to this table, we can also model a state machine with several types of diagrams. I'm partial to simple circles and lines, where the circles are the states and the lines are the transitions.

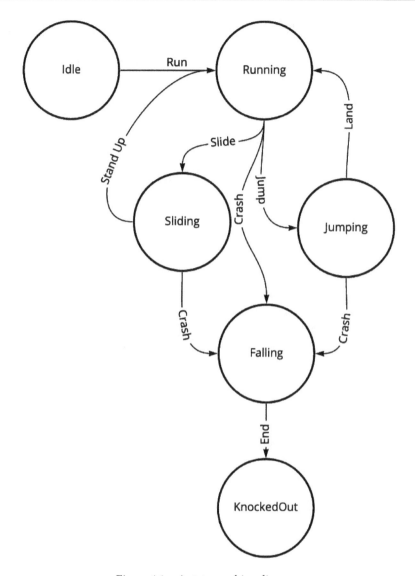

Figure 4.1 – A state machine diagram

This diagram is a fleshed-out version of the previous table, with all of the entries completed. It starts at the **Idle** state and transitions to the **Running** state via the **Run** event. From there, it can go in several directions. If the player *slides*, it can go into the **Sliding** state; if the player *jumps*, it can go into the **Jumping** state. Both of those eventually return to the **Running** state when sliding or jumping is over. **Running**, **Sliding**, and **Jumping** can all transition into the **Falling** state when they crash into something.

This does result in a lot of transitions across the middle of the diagram. Finally, the **Falling** state transitions into the **KnockedOut** state when **Falling** is over, via the **End** event. If you're familiar with this type of diagram, you might point out that I could have used a *superstate* to contain **Running**, **Jumping**, and **Sliding** and used one event to transition all of those to **Falling**. You'd be right, but we won't need to concern ourselves with that for our implementation.

You might be asking, what's the benefit of all this? Does this really fit the *minimal architecture* that we covered in the last chapter? Answering the second question first, the answer is, uh…maybe? I find that state machines help me keep code together that belongs together, rather than sprinkling `match` statements throughout my code base as I might have to when using a simple `enum`. That doesn't mean we won't have those `match` statements; they'll just be in one place.

I also find it fits my mental model of how code works well, and it helps prevent errors because you simply *can't* perform an action that's invalid because it's not available for that given state. Frankly, the state machine exists whether or not we model it, and it's cleaner if we can also model it in code rather than having it emerge accidentally. So, those are the benefits, and that's why I think it fits in our minimal architecture. Now it's time to implement it.

Implementing with types

The **Object-Oriented (OO)** state pattern is typically implemented as a variation on the strategy pattern, where you swap out different objects that all implement the same state interface at runtime based on the various transitions. The diagram looks something like this:

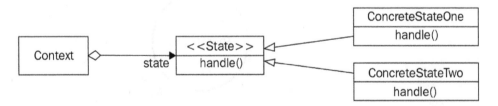

Figure 4.2 – State pattern

In the OO version of the pattern, **Context** has a reference to a **State** interface, and different states implement that interface. Frequently, the events, such as **handle**, take a reference to **Context** itself, in order to have side effects on the transitions. This design works reasonably well, but in Rust, we have two features that we'll use that differentiate it from the traditional pattern. The first is `enum`, which we can use to enumerate the states in a clearer fashion than traditional objects. The second is **generic types**, which we'll use to model each state as a **typestate**.

> **Important Note**
>
> The original state machine implementation I wrote was largely based on this excellent article by Ana Hobden, a.k.a. Hoverbear, at `https://hoverbear.org/blog/rust-state-machine-pattern/`. While this book no longer uses that pattern, I encourage you to read it for an alternative approach.

The typestate pattern

Typestate is a fancy name for embedding the state of an object in its type. The way it works is that you have a generic structure with one generic parameter representing the state. Then, each state will have methods that can return new states. So, instead of each state having common methods, as they do, as shown in *Figure 4.2*, each state has its own methods that return the new state. The states in *Figure 4.2* might look something like this:

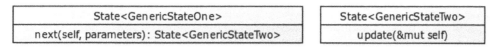

Figure 4.3 – Typestate pattern

In this diagram, `State<GenericStateOne>` has a `next` method, which consumes `self` and returns `State<GenericStateTwo>`. Meanwhile, `State<GenericStateTwo>` only has an `update` method, which takes a mutably borrowed `self`. The implications of this are that the compiler will catch you if you try to call `next` on `State<GenericStateTwo>`. In the traditional OO pattern, all states must handle all the same methods because they share an interface, so this kind of defense isn't possible. Often, this means implementing methods you don't actually care about, and either returning an error state or `Self`, and then debugging at runtime.

In addition, we can use the `mod` keyword and Rust's rules about privacy to make it impossible to create any state in an invalid state. We can make it impossible to move from `GenericStateOne` to `GenericStateTwo` without calling `next` by keeping the internals of `State` private so you can't just construct it. This is called **making illegal states unrepresentable**, and it's a great way to make sure you don't make mistakes in your programs.

> **Important Note**
>
> I tracked down the *making illegal states unrepresentable* phrasing to Yaron Minsky (`https://blog.janestreet.com/effective-ml-revisited/`); however, it's quite likely the practice and phrasing are older than that.

Typestates are intimidating because they are both a new concept and new jargon, so don't worry if you feel a little confused.

> **Tip**
>
> There's a lot of great information on typestates in Rust. There's an excellent talk by Will Crichton from Strange Loop (`https://youtu.be/bnnacleqg6k?t=2015`), as well as blogs at `https://docs.rust-embedded.org/book/static-guarantees/typestate-programming.htm` and `http://cliffle.com/blog/rust-typestate/`.

If you want to forget all about generics and type theory for a minute, they can be summarized as follows:

- Each state of the object is represented by a separate struct.

- You can only advance from one state to another by methods on that struct.

- You can guarantee you can only create valid states using privacy rules.

The rest are just details.

Finally, we're going to need an enum to *hold* our typestate. Each state is generic, so to continue in our preceding example, any struct that will interact with our state machine will need to hold *either* State<GenericStateOne> or State<GenericStateTwo>. In order to do that, we would either need to make the containing struct generic as well, and then create new versions of the containing struct every time the state changes, or wrap the generic object in an enum.

We'll use an enum because it prevents the generic nature of the typestate from propagating throughout the program, allowing the typestate to be an implementation detail. We're going to write the kind of state machine that Rust is very good at. Let's get to it.

Managing animation

We'll create our state machine to manage the different animations. Specifically, when RHB isn't moving, he's `Idle`, but when he's moving, he's `Running`. When he jumps, he's `Jumping`. You get the idea.

Those different RHB states correspond to the different animations managed using a state machine. We'll first create the RHB with a state machine and then integrate it into our current application. We'll implement this *top-down*, starting with a struct that represents RHB and letting the compiler errors drive further development. This is sometimes called **Compiler-Driven Development** although it's not a formalized approach such as **Test-Driven Development**. It can work extremely well in a language with a robust type system and great compiler errors, such as Rust. Let's start with how we'll represent RHB.

The `RedHatBoy` struct will contain the state machine, the sprite sheet, and the image because eventually, it will draw itself:

```
struct RedHatBoy {
    state_machine: RedHatBoyStateMachine,
    sprite_sheet: Sheet,
    image: HtmlImageElement,
}
```

> **Important Note**
>
> All of this code belongs in the `game` module. That means you can put it in the `game.rs` file or if you so choose, put it in a separate file and bring it into the `game` module with the `mod` keyword. I'll leave that up to you.

Of course, this won't work because you haven't created the state machine yet. You do have the `Sheet` structure from *Chapter 3, Creating a Game Loop*. Let's create `RedHatBoyStateMachine`:

```
#[derive(Copy, Clone)]
enum RedHatBoyStateMachine {
    Idle(RedHatBoyState<Idle>),
    Running(RedHatBoyState<Running>),
}
```

Seeing the enum we discussed earlier, it might still be unclear why we're using it when we'll be creating all of these typestate structures. RedHatBoyState, which doesn't exist yet, is a generic type that contains another type, where those types represent the various states. So, why the redundant enum? Because we want to be able to switch easily between the states without using the heap or dynamic dispatch. Let's imagine we defined the RedHatBoy struct in the following way:

```
struct RedHatBoy {
    state: RedHatBoyState<Idle>,
    sprite_sheet: Sheet,
}
```

Now the state is fixed to a state. We could, of course, define things in the following way:

```
struct RedHatBoy<T> {
    state: RedHatBoyState<T>,
    sprite_sheet: Sheet,
}
```

But of course, now RedHatBoy has to also be a generic type. You can make this work without the enum using Box<dyn State>, but that's not very ergonomic and it would require implementing the same methods on every state, so we'll stick with the enum. I have to acknowledge that I don't like the stutter in the types such as *Idle* (RedHatBoyState<*Idle*>), but we'll see that the enum wrapper becomes extremely useful as we implement the state machine. Make sure that the enum is Copy, Clone as well, for reasons you'll see shortly.

> **Important Note**
>
> If you're interested, *The Rust Programming Language* has a chapter that describes how to implement a state pattern in a traditional OO way. Interestingly, they eventually abandon it in favor of using an enum. You can find that here: https://bit.ly/3hBsVd4.

Of course, this code still doesn't compile, because we haven't created either of those states or the RedHatBoyState x. This is what I meant by Compiler-Driven Development. We can start by creating RedHatBoyState:

```
mod red_hat_boy_states {
    use crate::engine::Point;

```

```
#[derive(Copy, Clone)]
pub struct RedHatBoyState<S> {
    context: RedHatBoyContext,
    _state: S,
}

#[derive(Copy, Clone)]
pub struct RedHatBoyContext {
    frame: u8,
    position: Point,
    velocity: Point,
}
}
```

All the code relating to the individual states will go in its own module, red_hat_
boy_states, so that we can only make public the methods required by the rest of
the game module. This will make it impossible to accidentally create a state without
using the methods provided, and therefore, impossible to accidentally make an
invalid transition. The only way to transition from RedHatBoyState<Idle>
to RedHatBoyState<Running> is going to be through the methods on
RedHatBoyState<Idle>. It's important that both RedHatBoyState and
RedHatBoyContext are public but their members are private, so we can use them
as intended.

Inside the new module, RedHatBoyState is a simple generic type that contains
_state, which is never read, hence the underscore, and RedHatBoyContext. Now,
RedHatBoyContext is a structure with data that's common to all the states. In this
case, that's the frame being rendered, the position, and the velocity. We'll need it so that
the state transitions can modify the state of RHB. Putting all of this in the red_hat_
boy_states module means that we haven't changed the compiler error message. We
need to import that module into the game module with use self::red_hat_boy_
states::*;, which you can add anywhere in the game module. This gets us partway
there, but if we look at the following compiler output, we're still not finished:

```
error[E0412]: cannot find type 'Idle' in this scope
  --> src/game.rs:19:25
   |
19 |       Idle(RedHatBoyState<Idle>),
   |                              ^^^^ not found in
   this scope
```

There's also a corresponding enum variant for
Running(RedHatBoyState<Running>). Both Idle and Running don't exist. We
can create both of these easily, with empty structures inside the red_hat_boy_states
module. Note that both of these must also be Clone:

```
#[derive(Copy, Clone)]
struct Idle;
#[derive(Copy, Clone)]
struct Running;
```

Transitioning between states

Congratulations! You've created two states for RHB. This achieves…nothing. There's a bit
missing. For starters, we can't transition from Idle to Running, and those states don't
actually do anything when they aren't transitioning. Let's take care of a transition right
now. We'll add a method on RedHatBoyState<Idle> to go from Idle to Running:

```
mod red_hat_boy_states {
    ....
    impl RedHatBoyState<Idle> {
        pub fn run(self) -> RedHatBoyState<Running> {
            RedHatBoyState {
                context: self.context,
                _state: Running {},
            }
        }
    }
}
```

This is the transition from Idle to Running, and the run method is where the magic
happens. This is just a function that takes a RedHatBoy<Idle> state and converts it to
a RedHatBoy<Running> state, and for now, doesn't change any RedHatBoyContext
data. You might wonder then, what magic?

This means that to transition from Idle to Running, you can use run, but it also means
you can't transition from Running back into Idle, and that makes sense because the
game doesn't allow that behavior. The function also takes mut self, so that when it's
called, it consumes the current state. This means that if you want to somehow keep Idle
around after transitioning to Running, you have to clone it, and if you do that, you
probably really meant to do it.

You also can't create the Running state directly, because its data members are private, which means you can't just create that state by mistake. You can't create the Idle state either, and that's a problem because it's the start state. We'll address that in a moment, but first, let's dive into how we'll interact with the states through our state machine.

Managing the state machine

Initially, we might be tempted to implement our state machine by adding methods on the RedHatBoyStateMachine enum, as follows:

```
#[derive(Copy, Clone)]
enum RedHatBoyStateMachine {
    Idle(RedHatBoyState<Idle>),
    Running(RedHatBoyState<Running>),
}

impl RedHatBoyStateMachine {
    fn run(self) -> Self {
        match self {
            RedHatBoyStateMachine::Idle(state) =>
                RedHatBoyStateMachine::Running(state.run()),
            _ => self,
        }
    }
}
```

This isn't terrible, but it means that every method on our state machine will likely need to match the current variant of the RedHatBoyStateMachine enum. Then, it would return the new variant based on either the transition or self when the transition isn't currently valid. In other words, while the compiler will error if we call run on the Running state, it won't error if we call run on RedHatBoyStateMachine when the current variant is Running. This kind of error, where we call run by mistake on the wrong state, is exactly what we're trying to get away from with our typestates. We'd go to all the trouble of writing these typestates only to immediately throw away one of the benefits in every method on the RedHatBoyStateMachine enum.

Unfortunately, we can't completely get away from that problem, because we are using an enum to contain our states. There's no way to implement methods on variants of an enum as we can with generic structures, and if we're going to wrap the state in an enum, we'll have to match on the variant. What we can do is *reduce* the surface area of that kind of error by reducing the number of methods that operate in the states. Specifically, instead of calling run on the enum, we'll create a transition function that takes Event. That is going to look like the following code:

```
#[derive(Copy, Clone)]
enum RedHatBoyStateMachine {
    Idle(RedHatBoyState<Idle>),
    Running(RedHatBoyState<Running>),
}

pub enum Event {
    Run,
}

impl RedHatBoyStateMachine {
    fn transition(self, event: Event) -> Self {
        match (self, event) {
            (RedHatBoyStateMachine::Idle(state),
            Event::Run) => {
                RedHatBoyStateMachine::Running(state.run())
            }
            _ => self,
        }
    }
}
```

We've solved the problem caused by the enum with another enum! This is very *Rusty* of us. In this case, we've created an enum named Event to represent every event that could happen to our machine and replaced the method named run with a method named transition.

So, instead of many small methods for run, jump, and similar, we will have one method named `transition` and a bunch of `Event` variants. How does this improve things? Because there is only one `match` statement that we have to update when we want to add a transition, instead of potentially adding multiple little `match` statements. Keep in mind that this function takes `mut self`, which means calling `transition` will consume `self` and return a new `RedHatBoyStateMachine` just as the `run` method does on `RedHatBoyState<Idle>`.

Using Into for clean code

We can actually improve the ergonomics of this method using the `From` trait. If you're unfamiliar, the `From` trait is a Rust feature that lets us define how to convert from one type to another. Implementing the `From` trait on your type will also implement the `Into` trait, which will provide an `into` method that will make it easy to convert from one type to another.

We know that if we have `RedHatBoyState<Running>`, it will convert into the `RedHatBoyStateMachine::Running` variant, and if we write the conversion by implementing the `From` trait, we will be able to replace that wrapping with a call to `into`. That was a lot of words for a little bit of code, so the following is what the implementation of the `From` trait looks like:

```
impl From<RedHatBoyState<Running>> for RedHatBoyStateMachine {
    fn from(state: RedHatBoyState<Running>) -> Self {
        RedHatBoyStateMachine::Running(state)
    }
}
```

This can be placed right under the implementation of `RedHatBoyStateMachine`. It defines how to convert from `RedHatBoy<Running>` to `RedHatBoyStateMachine`, and it's the same small amount of code we wrote in the `transition` method. Because we have this now, we can make that method a little more succinct, as shown here:

```
impl RedHatBoyStateMachine {
    fn transition(self, event: Event) -> Self {
        match (self, event) {
            (RedHatBoyStateMachine::Idle(state),
            Event::Run) => state.run().into(),
            _ => self,
        }
    }
```

```
        }
    ...
```

Replacing calls like `RedHatBoyStateMachine::Idle::Running(state.run)` with `into` isn't just prettier and more concise; it also means that if `run` changes to return a different state, the `transition` method can stay the same, as long as a `From` trait has been written to go from the state to the `RedHatBoyStateMachine` enum. It's a nice little change that makes our code more flexible.

It's a little odd that the `RedHatBoyStateMachine` enum is what we call our state machine because we don't normally associate enumerated types with behavior, but this method is why we call it a machine. We use enum to hold the various generic states, and we use the ability to add methods to an enum to make it a lot more ergonomic to use. The various states know how to transition from one state to another, and the machine knows when to do the transitions.

Integrating the state machine

Now that we've built a state machine, albeit one with two states, we need to actually use it for something. Recall our current game, let RHB run throughout a meaningless void. We're going to want to change it so that RHB starts in the left corner and begins running when the user hits the *right arrow key*. In other words, they will transition from `Idle` to `Running`. When that happens, we'll also want to make sure we're showing the appropriate animation.

We'll start by putting `RedHatBoy` in the `WalkTheDog` game:

```
pub struct WalkTheDog {
    image: Option<HtmlImageElement>,
    sheet: Option<Sheet>,
    frame: u8,
    position: Point,
    rhb: Option<RedHatBoy>,
}

...

impl WalkTheDog {
    pub fn new() -> Self {
        WalkTheDog {
```

```
                image: None,
                sheet: None,
                frame: 0,
                position: Point { x: 0, y: 0 },
                rhb: None,
            }
        }
    }
```

RHB will need to be an `Option` for now because `RedHatBoy` contains a sprite sheet. Since the sprite sheet isn't available until the image is loaded in `initialize`, we have to make `rhb` an `Option` type. We'll want to initialize the machine in the `initialize` function, and for that purpose, we'll want to create a convenient `new` method for the `Idle` state:

```
mod red_hat_boy_states {
    use crate::engine::Point;
    const FLOOR: i16 = 475;
    ...
    impl RedHatBoyState<Idle> {
        pub fn new() -> Self {
            RedHatBoyState {
                context: RedHatBoyContext {
                    frame: 0,
                    position: Point { x: 0, y: FLOOR },
                    velocity: Point { x: 0, y: 0 },
                },
                _state: Idle {},
            }
        }
    }
    ...
```

Because `Idle` is the initial state, it's the only state that will get a `new` function, as mentioned earlier. We've also introduced a constant called `FLOOR` that marks the bottom of the screen, where RHB will land when he jumps.

I'll show it here as if it's defined right at the top of the `red_hat_boy_states` module. Now, in `Game` `initialize`, we still have a compiler error because we haven't set up `RedHatBoy` in the game. We can do that right after we've loaded the sprite sheet, and we'll keep two copies of the sprite sheet around; not because we want two copies, but because we'll delete all the old code when we've successfully replaced it with the new code. You can see the changes here:

```
#[async_trait(?Send)]
impl Game for WalkTheDog {
    async fn initialize(&self) -> Result<Box<dyn Game>> {
        let sheet: Option<Sheet> = browser::fetch_json(
        "rhb.json").await?.into_serde()?;
        let image = Some(engine::load_image(
        "rhb.png").await?);

        Ok(Box::new(WalkTheDog {
            image: image.clone(),
            sheet: sheet.clone(),
            frame: self.frame,
            position: self.position,
            rhb: Some(RedHatBoy::new(
                sheet.clone().ok_or_else(|| anyhow!
                    ("No Sheet Present"))?,
                image.clone().ok_or_else(|| anyhow!
                    ("No Image Present"))?,
            )),
        }))
    }
...
```

We had to change a surprising amount of code here, because of Rust's borrowing rules. Our intent is to `clone` `sheet` and `image` and send those into the `RedHatBoy::new` method. However, if we do that, we also need to clone `image` and `sheet` when setting the fields for `image` and `sheet` on `WalkTheDogStruct`. Why? Because the `image:` `image` line is a move, and can't be accessed after that. That's the borrow after move error. Instead we clone `image` and sheet and move the cloned instances into WalkTheDog. Then when creating the RedHatBoy we clone them again.

The same goes for `sheet`. We also have to explicitly call out the type of `sheet` when we assign it in the first place because the compiler can't infer the type anymore. Fortunately, this is an intermediate step; we are working past the compiler errors and will eventually reduce this code to what we actually need. We can't yet because we've replaced one compiler error with two!

Before, the `rhb` field wasn't filled in when we created `WalkTheDog`, so that didn't compile. In order to set the `rhb` field to something, we are presuming a `RedHatBoy::new` method exists, but it doesn't, so that doesn't compile. We are also passing the soon-to-exist constructor clones of `sheet` and `image`. The `Sheet` type doesn't support `clone` yet, so that doesn't compile either. We'll need to fix both of these compiler errors to move forward.

Before we continue, I want to note how we use the `ok_or_else` construct on each `clone` call, and then the `?` operator. `RedHatBoy` doesn't need to hold `Option<Sheet>` or `Option<HtmlImageElement>`, so its constructor will take `Sheet` and `HtmlImageElement`. Calling `ok_or_else` will convert `Option` into `Result`, and `?` will return from the `initialize` method with `Error` if the value isn't present. This prevents the rest of the code from having to continually validate that the `Option` type is present, so the code will be a little bit cleaner. The `Option` type is great, but at any time you can replace working with an `Option` type with the actual value it's wrapping.

The easiest of the two compiler errors to fix is the fact that `sheet` doesn't implement `clone`. Many in the Rust community derive `Clone` on any public type, and while I won't be following that practice in this book, there's no reason not to add it to `Sheet` and the types it references, as shown here. Remember, `Sheet` is in the `engine` module:

```rust
#[derive(Deserialize, Clone)]
pub struct SheetRect {
    pub x: i16,
    pub y: i16,
    pub w: i16,
    pub h: i16,
}

#[derive(Deserialize, Clone)]
pub struct Cell {
    pub frame: SheetRect,
}
```

```
#[derive(Deserialize, Clone)]
pub struct Sheet {
    pub frames: HashMap<String, Cell>,
}
```

Now, we're down to one compiler error, `RedHatBoy` doesn't have a `new` function, so let's create an `impl` block for the `RedHatBoy` struct and define that, as shown here:

```
impl RedHatBoy {
    fn new(sheet: Sheet, image: HtmlImageElement) -> Self {
        RedHatBoy {
            state_machine: RedHatBoyStateMachine::Idle(
              RedHatBoyState::new()),
            sprite_sheet: sheet,
            image,
        }
    }
}
```

This creates a new `RedHatBoy` with a state machine in the `Idle` state. We've also loaded `sprite_sheet` and `image` in the `initialize` function and passed them to this constructor. Congratulations! Our code compiles!

Drawing RedHatBoy

Unfortunately, this still doesn't do much. `RedHatBoy` is never drawn! The interface we want is to say `self.rhb.draw()` and see RHB drawing the idle animation. We also want to call the `run` function when we push the *right arrow* and see RHB run.

Let's start by implementing `draw` on `RedHatBoy`. We'll create a draw function that will mimic the draw function in `WalkTheDog` only using the shared `RedHatBoyContext` that's in `RedHatBoyState`. That code is as follows, written as part of the `impl` `RedHatBoy` block:

```
impl RedHatBoy {
    ...
    fn draw(&self, renderer: &Renderer) {
        let frame_name = format!(
            "{} ({}).png",
            self.state_machine.frame_name(),
```

```
                    (self.state_machine.context().frame / 3) + 1
        );

        let sprite = self
            .sprite_sheet
            .frames
            .get(&frame_name)
            .expect("Cell not found");

        renderer.draw_image(
            &self.image,
            &Rect {
                x: sprite.frame.x.into(),
                y: sprite.frame.y.into(),
                width: sprite.frame.w.into(),
                height: sprite.frame.h.into(),
            },
            &Rect {
                x: self.state_machine.context()
                  .position.x.into(),
                y: self.state_machine.context()
                  .position.y.into(),
                width: sprite.frame.w.into(),
                height: sprite.frame.h.into(),
            },
        );
    }
}
```

This is nearly identical to the code that exists in the draw function already for our
happily running RHB. Instead of always using the **Run** animation, now we're dynamically
choosing the animation based on the state of the system, via the frame_name function,
which doesn't exist yet.

We're also getting `position` and `frame` off `context()`, another function that doesn't exist yet. Again, we'll let the compiler guide us to create both of these functions; Compiler-Driven Development strikes again! The `RedHatBoyStateMachine` enum needs to provide a way to return `RedHatBoyContext` and `frame_name`. We can add those implementations, as follows:

```
impl RedHatBoyStateMachine {

    ...


    fn frame_name(&self) ->&str {
        match self {
            RedHatBoyStateMachine::Idle(state) =>
            state.frame_name(),
            RedHatBoyStateMachine::Running(state) =>
            state.frame_name(),
        }
    }


    fn context(&self) ->&RedHatBoyContext {
        match self {
            RedHatBoyStateMachine::Idle(state)
            =>&state.context(),
            RedHatBoyStateMachine::Running(state)
            =>&state.context(),
        }
    }
}
```

I admit I don't love either of these methods and did consider creating a trait that the various states would implement as an alternative. After some thought, I decided this was simpler, and because the Rust compiler will fail if you don't match every single enum variant, I'm willing to accept these duplicate *case* statements.

The `frame_name` and `context` methods both delegate to the currently active `state` to get the data that's required. In the case of `frame_name`, this will be a method that returns the name of the animation in `rhb.json` for a given state as defined on each state. The `context` method is particularly odd because we always return the same field for every single state and always will, as that data is shared across all the states. That's going to require a generic implementation, which we'll write in a moment. An exercise for you would be to simplify these functions with a macro, but we won't do that here.

> **Important Note**
>
> You might have noticed that the line `self.state_machine.context().position.x` violates the **Law of Demeter** and objects to it. The Law of Demeter is a style guideline for OO code that states that you should *only talk to your immediate friends*, and in this case, `self` should only talk to `state_machine` (its friend) but instead, it talks to `position` via `context`. This couples `RedHatBoy` to the internal structure of `RedHatBoyContext` in a way that could be avoided by adding getters for `position_x` and `position_y` on `state machine`, which would delegate to `context`, which would, in turn, delegate to `position`. The Law of Demeter is a great guideline when setting values, and you should almost always follow it for mutable data, but in this case the data is immutable. We can't change the context through this getter, and the downsides of violating the Law of Demeter are not as relevant. I don't feel it's necessary to create more delegating functions just to avoid violating an arbitrary guideline, but if it becomes a problem, we can always change it. For more information on this, go to `https://wiki.c2.com/?LawOfDemeter`.

Following the compiler again, we've moved the errors from the `draw` method on `RedHatBoy` into `RedHatBoyStateMachine` because none of the states have methods for `frame_name` or `context`. Out of these two methods, `frame_name` is more straightforward, so we'll implement it first. It's a getter of the name of the frame in the `rhb.json` file, and it's different for every state, so we'll put that method on every state, as shown here:

```
mod red_hat_boy_states {
    use crate::engine::Point;
    const FLOOR: i16 = 475;
    const IDLE_FRAME_NAME: &str = "Idle";
    const RUN_FRAME_NAME: &str = "Run";

    impl RedHatBoyState<Idle> {
```

```
        . . .
        pub fn frame_name(&self) -> &str {
            IDLE_FRAME_NAME
        }
    }
    . . .
    impl RedHatBoyState<Running> {
        pub fn frame_name(&self) -> &str {
            RUN_FRAME_NAME
        }
    }
}
```

We've added two constants, IDLE_FRAME_NAME and RUN_FRAME_NAME, which correspond to the names of the frames for the Idle and Run sections of our sprite sheets, respectively. We then created a new method, frame_name, on RedHatBoyState<Idle> as well as an entirely new implementation for RedHatBoyState<Running>, which also has a frame_name method.

It's worth thinking about whether we could use a trait object (https://bit.ly/3JSyoI9) instead of our enum for RedHatBoyStateMachine, and it probably is possible. I've experimented with it and didn't come to a satisfying solution, but I would encourage you to give it a shot. You'll learn a lot more from this book if you experiment with the code on your own.

Now that we've handled the frame_name method, we'll want to add a context method. That method is going to do the same thing for every state, return the context, and we can write it generically for all of them, as shown here:

```
mod red_hat_boy_states {
    . . . .
    #[derive(Copy, Clone)]
    pub struct RedHatBoyState<S> {
        context: RedHatBoyContext,
        _state: S,
    }

    impl<S> RedHatBoyState<S> {
        pub fn context(&self) -> &RedHatBoyContext {
```

```
            &self.context
        }
    }
...
```

This is a pretty cool feature of Rust. Since we have a generic struct, we can write methods on the generic type, and it will apply to all the types. Finally, there is one more compiler error, in the draw function where we reference the frame or position fields on context. These fields are private, but as long as RedHatBoyContext is an immutable type, we can each make of those public, as follows:

```
mod red_hat_boy_states {
    ...
    #[derive(Copy, Clone)]
    pub struct RedHatBoyContext {
        pub frame: u8,
        pub position: Point,
        pub velocity: Point,
    }
    ...
```

Finally, we need to call that method on RedHatBoy in the WalkTheDog#draw function. You can add that in this, admittedly awkward, one-liner right at the end of the draw function:

```
fn draw(&self, renderer: &Renderer) {
    ...

    self.rhb.as_ref().unwrap().draw(renderer);
```

If you've followed along successfully, you should see the following screen:

Figure 4.4 – RHBs

At the top we have our old, endlessly running RHB, and at the bottom our new RHB just standing still. The new version has fewer features; we've gone backward, but why? This prepared us for what we're going to do next, moving him around and changing animations. Speaking of animations, the `Idle` version of RHB isn't doing anything yet, because `frame` never changes. When RHB is idle, he stands while breathing slowly, so let's get that started, shall we?

Updating RHB

Our `RedHatBoy` struct is going to have an `update` function, which will, in turn, delegate to an `update` function on the state machine. It's a new method because every state is going to need to update, in order to advance the animation. We'll call `update` on `RedHatBoy` from `update` on `WalkTheDog`. That's a lot of updates, but it's really just delegation:

```
#[async_trait(?Send)]
impl Game for WalkTheDog {
    ...
    fn update(&mut self, keystate: &KeyState) {
....
        if self.frame < 23 {
            self.frame += 1;
        } else {
            self.frame = 0;
        }
        self.rhb.as_mut().unwrap().update();
```

```
        }
    }

impl RedHatBoy {
    ...
    fn update(&mut self) {
        self.state_machine = self.state_machine.update();
    }
}

impl RedHatBoyStateMachine {
    ...
    fn update(self) -> Self {
        match self {
            RedHatBoyStateMachine::Idle(mut state) => {
                if state.context.frame < 29 {
                    state.context.frame += 1;
                } else {
                    state.context.frame = 0;
                }
                RedHatBoyStateMachine::Idle(state)
            }
            RedHatBoyStateMachine::Running(_) => self,
        }
    }
}
```

In the update function on WalkTheDog, we've only added one new line, at the end of
the update function:

```
self.rhb.as_mut().unwrap().update();
```

It's funky because of the fact that rhb is Option, and we'll fix that in a little bit. We've added another small function to the RedHatBoy struct update that simply updates state_machine via the state machine's update function. This one line, and others like it, are why the state machine needs to be Copy. If it's not, then because update consumes self via the parameter of mut self, you'd have to use something like Option to move self into update, and then reset it again. By making everything Copy, you get a much more ergonomic update function.

Finally, the meat of the behavior is in the RedHatBoyStateMachine#update function. Here, we match on self and update the current frame on a mutable state parameter, and then return a new Idle state with a moved context with an updated frame. Unfortunately, this code doesn't compile; context isn't a public data member so you can't assign it. For now, we'll go ahead and make context public, but this should bother you. Remember that Law of Demeter guideline I mentioned earlier. It's one thing to get an immutable data value, another thing entirely to set a mutable value. This is the kind of coupling that could cause real problems down the line. We're *not* going to fix it right now, so go ahead and make context public, but we will be keeping a very close eye on this code.

At this point, if you look at update for WalkTheDog and update for RedHatBoyStateMachine, you'll see similarities. One is updating the running RHB in the upper left corner, and one is updating the idle RHB in the lower left. The time has come to begin combining these two objects. Let's go ahead and do that.

Adding the Running state

One thing to keep in mind about states is that they exist whether you implement a state machine or not. While we haven't implemented anything in RedHatBoyState<Running>, the Running state currently exists in WalkTheDog; RHB is running all around the void right now! We just need to move the details into our state machine, so that we as programmers can actually see the states and what they do as one coherent unit. Plus, then we'll stop having a sad and lonely boy who is running in place in the left-hand corner of the screen.

We can do that quickly by just modifying update in RedHatBoyStateMachine to match the version in Idle, with the different frame count for the run animation. That's shown as follows:

```
impl RedHatBoyStateMachine {
    ...
    fn update(self) -> Self {
        match self {
```

```
    . . .
            RedHatBoyStateMachine::Running(mut state) => {
                if state.context.frame < 23 {
                    state.context.frame += 1;
                } else {
                    state.context.frame = 0;
                }
                RedHatBoyStateMachine::Running(state)
            }
        }
    }
}
```

Now, the state machine is theoretically capable of drawing the run animation, but we haven't written anything to cause that transition. The other thing missing is potentially more subtle. The Running animation has 23 frames, and the Idle animation has 29. If we were to transform from Idle to Running with the frame count at 24, the game would crash.

Finally, I think we can all agree that the kind of duplication that we have here can be improved. The only difference between the two functions is the frame count. So, we have a few things to do:

1. Refactor the duplicated code.

    The code that updates context.frame suffers from a code smell called
    Feature Envy (https://bit.ly/3ytptHA) because the update function
    is operating over and over again on context. Why not move that function to
    RedHatBoyContext? That's shown here:

    ```
    const IDLE_FRAMES: u8 = 29;
    const RUNNING_FRAMES: u8 = 23;
    . . .
    impl RedHatBoyStateMachine {
        fn update(self) -> Self {
            match self {
                RedHatBoyStateMachine::Idle(mut state) => {
                    state.context =
                        state.context.update(IDLE_FRAMES);
                    RedHatBoyStateMachine::Idle(state)
    ```

```
            }
            RedHatBoyStateMachine::Running(mut state)
            => {
                state.context = state.context.update(
                RUNNING_FRAMES);
                RedHatBoyStateMachine::Running(state)
            }
        }
    }
}
mod red_hat_boy_states {
    ...
    impl RedHatBoyContext {
        pub fn update(mut self, frame_count: u8) ->
        Self {
            if self.frame < frame_count {
                self.frame += 1;
            } else {
                self.frame = 0;
            }
            self
        }
    }
}
```

RedHatBoyContext now has an update function that increments the
frame, looping it back to 0 when the total frame count is reached. Note how it
works the same way as our transitions, consuming self, and returning a new
RedHatBoyContext, although in reality, it's the same instance the entire time.
This gives us the same kind of *functional* interface that we're using elsewhere. The
total frame count changes with each state, so we pass that in as a parameter, using
constants for clarity.

2. Fix the Law of Demeter violation.

Looking at the two arms of each match statement, they are nearly identical, both mutating context in the way we didn't like earlier. Now is a good time to address it, which we can do by making the field private on RedHatBoyState<S> again, and creating new methods on the respective RedHatBoy state implementations, as shown here:

```
mod red_hat_boy_states {
    ...
    const IDLE_FRAMES: u8 = 29;
    const RUNNING_FRAMES: u8 = 23;
    ....
    impl RedHatBoyState<Idle> {
        ....
        pub fn update(&mut self) {
            self.context = self.context.update(
            IDLE_FRAMES);
        }
    }

    impl RedHatBoyState<Running> {
        ...
        pub fn update(&mut self) {
            self.context = self.context.update(
            RUNNING_FRAMES);
        }
    }
}
```

There! That's better. context is no longer inappropriately public, and each individual state handles its own updating. The only difference between them is the constant they use, and it's fitting to have that bundled with the implementation itself. Speaking of which, make sure you move the RUNNING_FRAMES and IDLE_FRAMES constants into the red_hat_boy_states module.

We'll need to modify the `update` method on `RedHatBoyStateMachine` to call this new method on each of the states:

```
impl RedHatBoyStateMachine {
    ....
    fn update(self) -> Self {
        match self {
            RedHatBoyStateMachine::Idle(mut state) =>
            {
                state.update();
                RedHatBoyStateMachine::Idle(state)
            }
            RedHatBoyStateMachine::Running(mut state)
            => {
                state.update();
                RedHatBoyStateMachine::Running(state)
            }
        }
    }
}
```

Each of the arms in update now updates the state, and then returns the state. There's some duplication here that's a little suspicious; we'll take another look at that shortly.

3. Move RHB on every `update`.

    If RHB is going to run in the running state, it needs to respect the velocity. In other words, update animates the frame, but it doesn't move, so let's add that to the `RedHatBoyContext` update method:

```
fn update(mut self, frame_count: u8) -> Self {
    ...

    self.position.x += self.velocity.x;
    self.position.y += self.velocity.y;
    self
}
```

Of course, RHB won't move yet because we aren't changing the velocity. That will come soon.

4. Ensure that the frame count resets to 0 when transitioning between states.

There are two categories of changes on the game object that can happen in our state machine. There are changes that happen when the state doesn't change. That's what update is and right now those are written in RedHatBoyStateMachine. There are also changes that happen on a transition, and those happen in the transition functions that are defined as methods of the type classes.

We already transitioned from Idle to Running via the run method, and we can make sure to reset the frame rate on the transition. That's a small change you can see here:

```
impl RedHatBoyContext {
    ...
    fn reset_frame(mut self) -> Self {
        self.frame = 0;
        self
    }
}
impl RedHatBoyState<Idle> {
    ....
    pub fn run(self) -> RedHatBoyState<Running> {
        RedHatBoyState {
            context: self.context.reset_frame(),
            _state: Running {},
        }
    }
}
```

RedHatBoyContext has grown a function called reset_frame, which resets its frame count to 0 and returns itself. By returning itself, we can chain calls together, which will come in handy shortly. The run method has also evolved to call reset_frame() on RedHatBoyContext and use that new version of context in the new RedHatBoyState struct.

5. Start Running on transition.

Now that we have prevented crashes by restarting animations on transitions, let's start running forward on a transition. This is going to be very short:

```
mod red_hat_boy_states {
    ....
```

```
const RUNNING_SPEED: i16 = 3;
...
impl RedHatBoyContext {
    ...
        fn run_right(mut self) -> Self {
            self.velocity.x += RUNNING_SPEED;
            self
        }
    }
    impl RedHatBoyState<Idle> {
        pub fn run(self) -> RedHatBoyState<Running> {
            RedHatBoyState {
                context: self.context.reset_frame()
                    .run_right(),
                _state: Running {},
            }
        }
    }
}
```

We've sprouted another method on `RedHatBoyContext` called `run_right`, which simply adds forward speed to the velocity. Meanwhile, we've chained a call (see!) to `run_right` in the transition. Don't forget to add the `RUNNING_SPEED` constant to the module.

6. Start Running on the *right* arrow.

Finally, we actually need to call this event when the `ArrowRight` button is pressed. At this point, we can follow along with where we're doing this in the `WalkTheDog` implementation:

```
impl Game for WalkTheDog {
    ...
    if keystate.is_pressed("ArrowRight") {
        velocity.x += 3;
        self.rhb.as_mut().unwrap().run_right();
    }
}

impl RedHatBoy {
```

```
        ...
    fn run_right(&mut self) {
        self.state = self.state.transition(
        Event::Run);
    }
}
```

This will now start our RHB running, so much so that he'll run right off the screen!

Figure 4.5 – This could be a problem

At this point, we could re-establish *moonwalking*, to bring RHB back on screen, but that doesn't really serve the purpose of the game. You can either create an event that resets horizontal velocity every update, just like the current code does, or you could track when a key goes up to remove some velocity. The second one feels better but will cause us to write a few events and possibly a transition from Running to Idle. No, we'll go to a third approach: ignore it and hit refresh! We don't need to move backward in our actual game, nor stop, so we won't. Let's not spend any more time writing code, that we'll just delete anyway. Speaking of that.

7. Delete the original code.

   Now that the new and improved RHB is moving, it's time to get rid of all the references in WalkTheDog to the sheet, the element, the frame...basically anything that isn't the RedHatBoy struct:

```
pub struct WalkTheDog {
    rhb: Option<RedHatBoy>,
}
```

Rather than boring you with endless deletes, I'll simply say you can delete all the fields that aren't rhb and follow the compiler errors to delete the rest of the code. When you're done, WalkTheDog becomes very short, as it should be. As for the arrow keys, you only need to worry about the ArrowRight key, and moving to the right.

> **Tip**
>
> As I said, we won't be restoring moving backward, up, or down here, but you
> could certainly consider restoring the walking backward functionality by
> extending the state machine. Doing so will help you internalize the lesson here
> and save you the trouble of refreshing all the time.

So, now RHB can run across the screen, but that's not much fun. Let's add sliding.

Transitioning to sliding

Transitioning from running to sliding will involve adding a new state for sliding, so that
we see the sliding action, but also checking for when a slide is complete and transitioning
back into the running state. This will mean sliding will have its own variation on the
update function. We can start by adding sliding on the *down* arrow and treating it all
just like running. We'll go through this quickly because most of it is familiar. Let's start by
adding sliding on the *down* arrow in the update method of WalkTheDog:

```
impl Game for WalkTheDog {
    fn update(&mut self, keystate: &KeyState) {
        ...

        if keystate.is_pressed("ArrowDown") {
            self.rhb.as_mut().unwrap().slide();
        }

    }
}
```

It's time to follow the compiler. RedHatBoy doesn't have a slide method, so let's add
that, as shown here:

```
impl RedHatBoy {
    ...
    fn slide(&mut self) {
        self.state_machine = self.state_machine.transition(
        Event::Slide);
    }
}
```

Transitioning via `Event::Slide` doesn't exist. There's no `Event::Slide` at all, so let's add those next:

```
enum Event {
    ....
    Slide,
}

impl RedHatBoyStateMachine {
    fn transition(self, event: Event) -> Self {
        match (self, event) {
            ...
            (RedHatBoyStateMachine::Running(state),
            Event::Slide) => state.slide().into(),
            _ => self,
        }
    }
    ...
```

There's nothing new in the preceding code block. When RHB is `Running`, it can transition to `Sliding` via the `Event::Slide` event and the `slide` method, which doesn't exist on the `RedHatBoyState<Running>` typestate. This is all very similar to how we went from `Idle` to `Running`.

To continue with the compiler, we need to add a `slide` method to the `RedHatBoyState<Running>` typestate, as in the following:

```
mod red_hat_boy_states {
    ...
    impl RedHatBoyState<Running> {
        ...
        pub fn slide(self) -> RedHatBoyState<Sliding> {
            RedHatBoyState {
                context: self.context.reset_frame(),
                _state: Sliding {},
            }
        }
    }
}
```

The `slide` method on `RedHatBoyState<Running>` converts the state into
`RedHatBoyState<Sliding>`, only calling `reset_frame` on `context` to
make sure the sliding animation starts playing at frame `0`. We also call `into` on the
`slide` method, which needs to convert `RedHatBoyState<Sliding>` into a
`RedHatBoyStateMachine` variant. That means we need to create the variant and
create a `From` implementation for it, as shown here:

```
enum RedHatBoyStateMachine {
    . . .
    Sliding(RedHatBoyState<Sliding>),
}
impl From<RedHatBoyState<Sliding>> for RedHatBoyStateMachine {
    fn from(state: RedHatBoyState<Sliding>) -> Self {
        RedHatBoyStateMachine::Sliding(state)
    }
}
```

At this point, you'll see errors on the `frame_name`, `context`, and `update` methods
of `RedHatBoyStateMachine` because their corresponding `match` calls don't have
cases for the new `Sliding` variant. We can fix that by adding cases to those `match`
statements, which will mimic the other cases:

```
impl RedHatBoyStateMachine {
    . . .
    fn frame_name(&self) -> &str {
        match self {
            . . .
            RedHatBoyStateMachine::Sliding(state) =>
            state.frame_name(),
        }
    }

    fn context(&self) ->&RedHatBoyContext{
        match self {
            . . .
            RedHatBoyStateMachine::Sliding(state)
            => &state.context(),
        }
```

```
        }

    fn update(self) -> Self {
        match self {
            RedHatBoyStateMachine::Sliding(mut state) => {
                state.update();
                RedHatBoyStateMachine::Sliding(state)
            }
        }
    }
}
```

Once again, we've replaced one compiler error with another. There is no `Sliding` state, and it doesn't have the methods we assumed it would. We can fix that by filling it in, adding some constants for good measure:

```
mod red_hat_boy_states {
    const SLIDING_FRAMES: u8 = 14;
    const SLIDING_FRAME_NAME: &str = "Slide";
    ...
    #[derive(Copy, Clone)]
    struct Sliding;

    impl RedHatBoyState<Sliding> {
        pub fn frame_name(&self) -> &str {
            SLIDING_FRAME_NAME
        }

        pub fn update(&mut self) {
            self.context = self.context.update(
            SLIDING_FRAMES);
        }
    }
}
```

If you look through this code, you'll see it's very similar to our already existing running code. If you followed along, you'll see RHB start skidding across the floor until he goes past the right edge of the screen:

Figure 4.6 – Safe

Stopping RHB from sliding is a little different than what we've done before. What we need to do is identify when the slide animation is complete, then transition right back into running without any user input. We'll start by checking whether the animation is done in the update method of the enum, which represents our machine, and then create a new transition from sliding back into running. We can do that by modifying the RedHatBoyStateMachine update method to check after updating in the sliding branch, as follows:

```
fn update(self) -> Self {
    match self {
        ...
        RedHatBoyStateMachine::Sliding(mut state) => {
            state.update(SLIDING_FRAMES);
            if state.context().frame >= SLIDING_FRAMES {
                RedHatBoyStateMachine::Running(
                    state.stand())
            } else {
                RedHatBoyStateMachine::Sliding(state)
            }
        }
    }
}
```

This doesn't compile yet, because stand isn't defined yet and because SLIDING_FRAMES is in the red_hat_boy_states module. You might think that we can make SLIDING_FRAMES public and define a stand method, or we could move SLIDING_FRAMES into the game module. These will both work but I think it's time to look a little more holistically at our update method.

Every arm of the `match` statement updates the current state and then returns a new
state. In the case of `Running` and `Idle`, it was always the same state, but in the case of
`Sliding`, sometimes it's the `Running` state. It turns out `update` is a transition, just one
that sometimes transitions to the state it started from. In a state diagram, it looks like this:

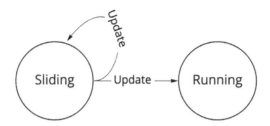

Figure 4.7 – Sliding to Running

If we wanted to be strict about it, we could say that **Sliding** transitions to an `Updating`
state when it gets an **Update** event, then it can transition back to **Sliding** or **Running**. This
is a case where the state exists, at least conceptually, but we don't actually have to create it
in our code.

`update` on the `Sliding` state is really best modeled as a transition because it's a method
that ultimately returns a state. Come to think of it, that's exactly what the other arms in
the `update` method are too! Yes, they don't ever transition to another state, but each
branch calls `update` and then returns a state. So, before we add `Sliding` to the `update`
method, let's refactor to make `update` a transition for both of the other states.

Since we're using Compiler-Driven Development, we'll change the `update` method to
work as if `update` is already a transition:

```
pub enum Event {
    ...
    Update,
}

impl RedHatBoyStateMachine {
    fn transition(self, event: Event) -> Self {
        match (self, event) {
            (RedHatBoyStateMachine::Idle(state),
            Event::Run) => state.run().into(),
            (RedHatBoyStateMachine::Running(state),
            Event::Slide) => state.slide().into(),
            (RedHatBoyStateMachine::Idle(state),
```

```
                    Event::Update) => state.update().into(),
                (RedHatBoyStateMachine::Running(state),
                    Event::Update) => state.update().into(),
                _ => self,
            }
        }
        ...

    fn update(self) -> Self {
        self.transition(Event::Update)
    }
}
```

With these changes, we've turned `Update` into `Event` and added two more arms to `match` in the `transition` method. Both of those arms work the same way as the other transitions: they call a method on the typestate and then convert the state into the `RedHatBoyStateMachine` enum with the `From` trait. The compiler error you get now might be a little strange; it looks like this:

```
error[E0277]: the trait bound 'RedHatBoyStateMachine: From<()>'
is not satisfied
   --> src/game.rs:155:83
    |
155 |                 (RedHatBoyStateMachine::Idle(state),
Event::Update) => state.update().into(),
    |                                                  |
    ^^^^ the trait 'From<()>' is not implemented for
'RedHatBoyStateMachine'
```

You may have expected that the error would say something about the `update` method not returning anything, but remember all Rust functions return something; they just return `Unit` when they don't return anything else. So, this error is telling you there's no way to convert from the `()`, or `Unit`, to a value of the `RedHatBoyStateMachine` type. That's not what we want to fix; we want to make both of the `update` calls on the states return new states. Those changes are next:

```
mod red_hat_boy_states {
    impl RedHatBoyState<Idle> {
        ...
        pub fn update(mut self) -> Self {
            self.context = self.context.update(
```

```
            IDLE_FRAMES);
        self
    }
}
...
impl RedHatBoyState<Running> {
    ...
    pub fn update(mut self) -> Self {
        self.context = self.context.update(RUNNING_FRAMES);
        self
    }
}
...
```

The changes are small but important. The update method for RedHatBoyState<Idle> and RedHatBoyState<Running> both return Self now, because even though the state doesn't change, these are still typestate methods that return a new state. They also take mut self now instead of &mut self. You can't return self if you mutably borrow it, so this method stopped compiling. More importantly, this means these methods don't make unnecessary copies. They take ownership of self when called, and then return it. So, if you're worried about an optimization problem because of extra copies, you don't have to be.

Now, we're down to one compiler error, which we've seen before:

```
the trait 'From<red_hat_boy_states::RedHatBoyState<red_hat_boy_
states::Idle>>' is not implemented for 'RedHatBoyStateMachine'
```

We didn't implement a conversion from the Idle state back to the RedHatBoyStateMachine enum. That's similar to the other ones we wrote, implementing From<RedHatBoyState<Idle>>, as shown here:

```
impl From<RedHatBoyState<Idle>> for RedHatBoyStateMachine {
    fn from(state: RedHatBoyState<Idle>) -> Self {
        RedHatBoyStateMachine::Idle(state)
    }
}
```

Remember that these implementations of the From trait are not in the red_hat_boy_states module. The red_hat_boy_states module knows about the individual states but does not know about RedHatBoyStateMachine. That's not its job.

Now that we've refactored the code, our little RHB doesn't slide anymore. Instead, he kind of sits down because the `Sliding` state doesn't handle the `Update` event. Let's fix that now.

Transitioning to sliding and back again

Part of the reason we used the typestate pattern for our individual states is so that we get compiler errors when we make a mistake. For instance, if we call `run` when we are in the `Running` state, it won't even compile because there is no such method. There is one place this doesn't hold, the `transition` method on the `RedHatBoyStateMachine` enum. If you call `transition` with a `RedHatBoyStateMachine` variant and an `Event` variant pair that don't have a match, it returns `Self`.

That's why our RHB is sitting down. He transitions to `Sliding` and then stops updating, staying in the same state forever. We'll fix that by adding the match for the `Update` event and then, you guessed it, follow the compiler to implement the sliding animation.

This starts by adding the match to the transition method, as shown here:

```
impl RedHatBoyStateMachine {
    fn transition(self, event: Event) -> Self {
        match (self, event) {
            ...
            (RedHatBoyStateMachine::Sliding(state),
             Event::Update) => state.update().into(),
            _ => self,
        }
    }
}
```

This match is just like the others; we match on `Sliding` and `Update` and call `update`. Just like before, we'll get an error:

```
the trait 'From<()>' is not implemented for
'RedHatBoyStateMachine'
```

The `Sliding` state still has an update method that doesn't return a state. That's not going to work with our current setup, but it's not as simple as making the `update` method return `Self`, as on the other two states.

Remember, there are two possible states that can come from the update method on Sliding: Sliding and Running. How is that going to work with our current setup? What we'll need to do is have update return an SlidingEndState enum that can be either Sliding or Running, and then we'll implement a From trait that will convert that into the appropriate variant of RedHatBoyStateMachine. That's odd to explain, so let's see it in action. We can modify the update method on RedHatBoyState<Sliding> to work like the one we proposed at the beginning of this section:

```
mod red_hat_boy_states {
    ...
    impl RedHatBoyState<Sliding> {
        ...
        pub fn update(mut self) -> SlidingEndState {
            self.context = self.context.update(
            SLIDING_FRAMES);

            if self.context.frame >= SLIDING_FRAMES {
                SlidingEndState::Complete(self.stand())
            } else {
                SlidingEndState::Sliding(self)
            }
        }
    }
}
```

We've taken the code that we originally considered putting in the RedHatBoyStateMachine update method and moved it into the update method of RedHatBoyState<Sliding>. This makes sense conceptually; the state should know how it behaves. On every update, we update context, and then check whether the animation is complete, with if self.context.frame >= SLIDING_FRAMES. If the animation is complete, we return one variant of this new enum that doesn't exist yet: SlidingState. The SlidingState variant can either be Complete or Sliding.

> **Important Note**
>
> It's definitely a little strange that the `update` method doesn't return another state here, and probably means we aren't using a *pure* typestate method. An alternative might have been to return the next `Event` from `update` and send that back into a call to the `transition` method on `RedHatBoyStateMachine`. That implementation ends up looking very strange because states are returning `Event`s that are only used by `RedHatBoyStateMachine` and are otherwise unreferenced in the red_ hat_boy_states module. Regardless of whether the strange return value of `update` makes you uncomfortable, I would encourage you to try other approaches. Maybe yours is better than mine!

Following the compiler yet again, we have two obvious problems: there is no `stand` method and there is no `SlidingEndState` enum. We can handle both of these right here, next to the code we just wrote, as shown:

```
impl RedHatBoyState<Sliding> {
    ...
    pub fn stand(self) -> RedHatBoyState<Running> {
        RedHatBoyState {
            context: self.context.reset_frame(),
            _state: Running,
        }
    }
}
pub enum SlidingEndState {
    Complete(RedHatBoyState<Running>),
    Sliding(RedHatBoyState<Sliding>),
}
```

The only side effect of the transition to `Running` is that we call `reset_frame` again on `context`. Remember this has to be done on every transition, otherwise, the program can try to animate the new state with `frame`, which isn't valid and will crash. So, we'll reset the frame back to `0` on every transition.

This leaves us with a compiler error to fix once again. This time, it's the following:

```
the trait 'From<SlidingEndState>' is not implemented for
'RedHatBoyStateMachine'
```

Pay close attention to that source trait. It's not coming from one of the states but from the intermediate `SlidingEndState`. We'll solve it the same way as before, with a `From` trait, but we'll need to use a `match` statement to pull it out of the `enum`:

```
impl From<SlidingEndState> for RedHatBoyStateMachine {
    fn from(end_state: SlidingEndState) -> Self {
        match end_state {
            SlidingEndState::Complete(running_state) =>
            running_state.into(),
            SlidingEndState::Sliding(sliding_state) =>
            sliding_state.into(),
        }
    }
}
```

Here, we match on `end_state` to get the actual `State` out of enum, and then call `into` on that state again to get to `RedHatBoyStateMachine`. A little boilerplate, but it makes it easier to do the conversion.

And now we have it! Run the game now and you'll see RHB take a short slide and pop back up again to the running state. Now that we've added three animations, it's time to deal with these ugly lines in the `WalkTheDog` implementation: `self.rhb.as_mut().unwrap().slide()`.

We treat `rhb` as an `Option` type, not because it's ever really going to be `None`, but because we don't have it yet before the `WalkTheDog` struct is initialized. After `WalkTheDog` is initialized, `rhb` can never be `None` again because the state of the system has changed. Fortunately, we now have a tool for dealing with that, the good old state machine!

Every little thing I think I see

WalkTheDog can be in two states, Loading or Loaded, after it's initialized. Fortunately, we accounted for this when we wrote our GameLoop. Remember that GameLoop returns Result<Game> from initialize; we're just currently always returning Ok(WalkTheDog). What if we made WalkTheDog an enum and returned a different state of our game instead? That would mean WalkTheDog would be a state machine, with two states, and initialize would become the transition! That's exactly what we're going to do. Modify WalkTheDog so it is no longer a struct but an enum, as shown here:

```
pub enum WalkTheDog {
    Loading,
    Loaded(RedHatBoy),
}
```

This is great; now everything is broken! Whoops! We'll need to adjust the WalkTheDog implementation to account for the two variants. First, we'll change the initialize function on WalkTheDog:

```
#[async_trait(?Send)]
impl Game for WalkTheDog {
    async fn initialize(&self) -> Result<Box<dyn Game>> {
        match self {
            WalkTheDog::Loading => {
                let json = browser::fetch_json(
                    "rhb.json").await?;

                let rhb = RedHatBoy::new(
                    json.into_serde::<Sheet>()?,
                    engine::load_image("rhb.png").await?,
                );
                Ok(Box::new(WalkTheDog::Loaded(rhb)))
            }
            WalkTheDog::Loaded(_) => Err(anyhow!
                ("Error: Game is already initialized!")),
        }
    }
    ...
```

Remember in *Chapter 3*, *Creating a Game Loop*, where we made this function return Game? This was why! In order to ensure `initialize` is only called once, `initialize` has to match `self` on its variants, and if we call `initialize` twice, we'll return an error via `anyhow!`. Otherwise, everything inside the `Loading` branch is the same as before, except we return `WalkTheDog::Loaded` instead of `WalkTheDog`. This does cause a compiler warning, which will become an error in future versions of Rust because `RedHatBoy` isn't public but is exposed in a public type. To get rid of that warning, you'll need to make `RedHatBoy` public, and that's fine; go ahead and do that. We also need to change the `new` constructor to reflect the new type, as shown here:

```
impl WalkTheDog {
    pub fn new() -> Self {
        WalkTheDog::Loading
    }
}
```

The `WalkTheDog` enum starts in `Loading`, nothing fancy there. The `update` and `draw` functions now both need to reflect the changing states; you can see those changes here:

```
#[async_trait(?Send)]
impl Game for WalkTheDog {
    ...
    fn update(&mut self, keystate: &KeyState) {
        if let WalkTheDog::Loaded(rhb) = self {
            if keystate.is_pressed("ArrowRight") {
                rhb.run_right();
            }

            if keystate.is_pressed("ArrowDown") {
                rhb.slide();
            }

            rhb.update();
        }
    }

    fn draw(&self, renderer: &Renderer) {
        ...
```

```
        if let WalkTheDog::Loaded(rhb) = self {
            rhb.draw(renderer);
        }
    }
}
```

You could argue this isn't really a change on the `Option` type, as we still need to check the state of `Game` each time we operate on `rhb`, and that's true, but I think this more clearly reveals the intent of the system. It also has the benefit of getting rid of the `as_ref`, `as_ mut` code, which is often confusing. Now that we've cleaned up that code, let's add one more animation to RHB. Let's see this boy jump!

Transitioning to jumping

Going through each and every change yet again for the jump is redundant. Instead, I can recommend you make the following change:

```
impl Game for WalkTheDog {
    ...
    fn update(&mut self, keystate: &KeyState) {
        if let WalkTheDog::Loaded(rhb) = self {
            ...
            if keystate.is_pressed("Space") {
                rhb.jump();
            }
        }
    }
}


impl RedHatBoy {
    ...
    fn jump(&mut self) {
        self.state_machine = self.state_machine.transition(
        Event::Jump);
    }
}
```

You should be able to follow the compiler errors all the way through, creating a transition from `Running` to `Jumping`. You can also look up the constant values you need directly out of `rhb.json`. The number of frames is the number of images in `Jump` in the animation multiplied by 3, and subtracting 1, and the name of the animation is `Jump`. Make sure you handle the `update` event in the transition method for `Jumping`.

Do all that and you see RHB skidding across the ground, doing a kind of dance:

Figure 4.8 – That's...not jumping

> **Tip**
>
> If you get stuck, the answers to this are available at `https://github.com/PacktPublishing/Rust-Game-Development-with-WebAssembly/tree/chapter_4/`. However, I strongly recommend trying to do this without checking first. Look at what we did for the first three transitions and try to understand what we did. Even if you get stuck, the time spent practicing is valuable here.

If you've correctly implemented the code for transitioning to the jumping state, our RHB will play his jumping animation, forever, while skidding across the ground. We've seen this before with the slide state, so it's time to figure out what's different about jumping. Of course, we know exactly what's different about jumping – you go up! Well, at least a little.

There are three things we need to do. First, we give RHB vertical velocity when he jumps; second, we need to add gravity so that RHB will actually come down when he jumps. And finally, we need to transition running when we land, using our ever-durable state machine:

1. Going up on `Jump`.

    Take a moment and think, where does this belong? Should it go in the `update` function, the `jump` event, or maybe in the `enum` implementation? No, this is a transition change because it happens on `jump`, and it belongs in the `jump` method on the `Running` type class. You should already have a transition from running to jumping, so let's update that function to add vertical velocity:

    ```
    mod red_hat_boy_states {
        ...
    ```

```
const JUMP_SPEED: i16 = -25;
...


impl RedHatBoyState<Running> {

    ...
    pub fn jump(self) -> RedHatBoyState<Jumping> {
        RedHatBoyState {
            context: self.context.set_vertical_
                velocity(JUMP_SPEED).reset_frame(),
            _state: Jumping {},
        }
    }
    ...
impl RedHatBoyContext {

    ...
    fn set_vertical_velocity(mut self, y: i16) ->
        Self {
            self.velocity.y = y;
            self
    }
```

Remember in our 2D coordinate system, y is 0 at the top, so we need a negative velocity to go up. It also resets the frame so that the jump animation starts at frame 0. The implementation in RedHatBoyContext is using the same pattern of accepting mut self and returning a new RedHatBoyContext. Now, if you let the app refresh, RHB will take off like Superman!

2. Adding gravity.

In order to have a natural jump, we'll apply gravity on every update. We'll do this *regardless of state* because later, we'll need to have RHB fall off of platforms and cliffs, and we don't want to have to constantly pick and choose when we're applying gravity. This will go in the update function of RedHatBoyContext, right at the top:

```
mod red_hat_boy_states {
...
const GRAVITY: i16 = 1;


    impl RedHatBoyContext {
```

```
fn update(mut self, frame_count: u8) -> Self {
    self.velocity.y += GRAVITY;
```

If you refresh the page right now, you'll get a blink-and-you'll-miss-it problem, and you'll probably be greeted with a blank screen. The screen isn't really blank; RHB just fell right through the ground!

Figure 4.9 – Tell my family I love them

We'll need to address this with our first case of **collision resolution**.

3. Landing on the ground.

    This is a bit of a spoiler for the next chapter, but collision detection happens in two steps. The first is detection, finding places where things collide, and the second is resolution, where you do something about the collision. Since there isn't anything to collide with in RHB's empty void, we can just do a simple check in the same update function to see whether his new position is past the floor and update the position back to the floor. Keep in mind, you do this *after* you update to a new position:

```
impl RedHatBoyContext {
    pub fn update(mut self, frame_count: u8) ->
    Self {
        ...
        self.position.x += self.velocity.x;
        self.position.y += self.velocity.y;

        if self.position.y > FLOOR {
            self.position.y = FLOOR;
        }
```

This may feel redundant, but we can't know gravity pulled RHB past the ground without actually calculating where he ends up, and we don't draw the in-between state, so the performance cost is minimal. This change prevents RHB from falling through the ground and causes a nice jumping arc, but he keeps performing the jumping animation for eternity. We need to change the state from `Jumping` back to `Running`, and we need to make that decision in `RedHatBoyStateMachine` because it's a conditional state change based on a condition just like the one that transitioned from `Sliding` to `Running`.

That's a change to the state machine, much like the one we did for `Sliding`, as seen here:

```
impl RedHatBoyState<Jumping> {
    ...
    pub fn update(mut self) -> JumpingEndState {
        self.context = self.context.update(
        JUMPING_FRAMES);
        if self.context.position.y >= FLOOR {
            JumpingEndState::Complete(self.land())
        } else {
            JumpingEndState::Jumping(self)
        }
    }
}
```

So, if the position is on the floor, we need to transition to `Running` via the `stand` method, only we can't! We never wrote a transition from `Sliding` to `Running`, just the other way around. We also never wrote a `JumpingEndState` enum, or a way to convert out of it via `From`. So, right now, you should see several compiler errors about all of that, the first being the following:

```
error[E0599]: no method named 'land' found for struct
'red_hat_boy_states::RedHatBoyState' in the current scope
   --> src/game.rs:413:48
    |
258 |     pub struct RedHatBoyState<S> {
    |     -------------------------- method 'land' not
found for this
```

There's the compiler error, but there's no `land` method. So, go write it. *I'm serious: go write it yourself. I'm not going to reproduce it here.* You can go ahead and follow along with the previous methods we wrote and implement them. You can do it; I believe in you. When you do, you'll have a clean animation from `Idle` to `Running`, then `Jumping`, and back to `Running` again. Then, you'll wander off the screen because we don't have a full scene yet, but we're getting there!

> **Important Note**
>
> If you get stumped, you can always check the source code for this chapter in the repository at `https://github.com/PacktPublishing/Rust-Game-Development-with-WebAssembly/tree/chapter_4/`.

Summary

This chapter covered one topic, but one of the most important topics in game development. State machines are everywhere in games, which we saw when we implemented a small one to manage the `Loaded` and `Loading` states of the `WalkTheDog` enum itself. They are a particularly nice way to implement animation states that must correspond with what the player is doing, and Rust has great ways to implement this pattern. We used two: the simple one for `WalkTheDog`, and the much more complex `RedHatBoyStateMachine` that uses the typestate pattern. The typestate pattern is a commonly used pattern in Rust, both inside and outside of game development, so you can expect to see it in many Rust projects.

We also used the compiler to drive development, over and over again. It's an incredibly useful technique, where you can start with what you want the code to look like and use the compiler's error messages to help you fill in the rest of the implementation. The code becomes like a paint by numbers picture, where you use higher-level code to draw the lines and the compiler error messages tell you how to fill them in. Rust has very good compiler error messages, getting better with every release, and it will pay huge dividends for you to pay close attention to them.

Now that our RHB can run and jump, how about he runs and jumps on something? We'll put him in a scene and have him jump on it in the next chapter.

5
Collision Detection

To make our game fun, our little **Red Hat Boy** (**RHB**) needs to run, jump, and slide. Fortunately, we just implemented all that, but he also needs to have something to jump on, something to slide under, and something to crash into. To make this game fun, we'll need to add **collision detection**, which is one of the most fun and most complicated aspects of game design.

Collision detection begins with math, detecting whether or not two shapes intersect, but leads to all kinds of interesting questions. We'll deal with some of those in this chapter, such as, how do we handle transparency in sprites? What do we do to make sure a player lands on a platform from above but crashes into a platform if they're underneath it? What about sprites that have shapes that aren't a simple box? It's going to be a blast!

In this chapter, we will cover the following topics:

- Creating a real scene
- Axis-aligned bounding boxes
- Getting bounding boxes from the sprite sheet
- Crashing into a stone
- Landing on and falling off a platform

By the end of this chapter, you'll have a real game, although it will be a short one. You'll have the skills to build your own scenes with good-looking collision detection, and you'll know how to integrate collision events with your own programs. You'll be able to, if you want, add your own new objects to the scene and crash into them or jump off them, or even fall off the world. Let's get started!

Technical requirements

You'll need to download the latest assets for this chapter from `https://github.com/PacktPublishing/Game-Development-with-Rust-and-WebAssembly/wiki/Assets`.

You can get the source code for this chapter at `https://github.com/PacktPublishing/Game-Development-with-Rust-and-WebAssembly/tree/chapter_5`.

No new assets are in the download, so if you downloaded them earlier, you don't need to do it again.

Check out the following video to see the Code in Action: `https://bit.ly/36BJYJd`

Creating a real scene

At the moment, RHB can move anywhere he wants, in an empty void, such as the one in *The Matrix*. It's progress; all that animation was real work, but it's not a game. It's time we put RHB in a setting – a background, platforms, maybe something to jump over. Let's start with a background.

Adding the background

Right now, our game can only render images from a sprite sheet, which we can use for a background, but that's overkill for one image. Instead, we'll add a new `struct` that draws a simple image from a `.png` file. Then, we'll add that to the `draw` and `initialize` functions in `WalkTheDog`:

1. Create an `Image` struct.

   We can work bottom-up for these changes, adding code to the engine and then integrating it into the game. Our `Image` `struct` will use a lot of the same code that we wrote in *Chapter 2, Drawing Sprites*, but with a simpler setup because we won't be using a sheet. All of this code should go into the `engine` module.

Start with a `struct` holding `HtmlImageElement`:

```
pub struct Image {
    element: HtmlImageElement,
    position: Point,
}

impl Image {
    pub fn new(element: HtmlImageElement, position:
        Point) -> Self {
        Self { element, position }
    }
}
```

There's nothing here you haven't seen before in another form. The `Image struct` holds the image element, presumably loaded via the `load_image` function, and its position in the scene. `Image` will also need a draw function, but there's no simple way to draw the entire image as it is in `Renderer`. That will need a new method, as shown here:

```
impl Renderer {
    ...

    pub fn draw_entire_image(&self, image:
        &HtmlImageElement, position: &Point)
            self.context
            .draw_image_with_html_image_element(image,
                position.x.into(), position.y.into())
            .expect("Drawing is throwing exceptions!
                Unrecoverable error.");
    }
}
```

This function is very similar to the `draw_image` function we wrote earlier, but it's using the simpler version of the JavaScript `drawImage` function that only takes an image and a position. To use this method, you'll need to be aware of how large the image you're drawing is. If it's too big or too small, it will show up just as big or small as the source image.

2. Now that you've added a method to `Renderer`, go ahead and update the `Image` implementation to draw an image with it:

```
impl HtmlImageElement {

    ...

    pub fn draw(&self, renderer: &Renderer) {
        renderer.draw_entire_image
            (&self.element,&self.position)
    }
}
```

Now that you can draw an image, let's load it.

3. Load the image.

The background image can be found in the downloaded assets, in `original/freetileset/png/BG/BG.png`, and can be copied into the `static` directory. Then, it can be loaded and used to create an `Image` struct. That will be done in the game module, in the `initialize` function of `WalkTheDog`, as shown here:

```
impl Game for WalkTheDog {
    async fn initialize(&mut self) -> Result<Box<dyn
Game>> {
    match self {
        WalkTheDog::Loading => {
            let sheet = browser::fetch_json
                ("rhb.json").await?.into_serde()?;
            let background = engine::
                load_image("BG.png").await?;

        ....
```

In the preceding code snippet, only the highlighted last line is new, which loads the background from a file. Our `WalkTheDog` enum only holds `RedHatBoy`, so we're going to have to restructure the code a little. While we could have the `WalkTheDog::Loaded` state hold a tuple of `RedHatBoy` and `Background`, that's going to get real annoying, real fast.

4. To do that, change `enum` to look like this:

```
pub enum WalkTheDog {
    Loading,
    Loaded(Walk),
}
```

We'll have `WalkTheDog` represent our game, but I decided that RHB takes the dog for "**walks,**" so our level will be `Walk`. In a generic framework, I might call this a "**scene**" or "**level,**" but this is a specific game, so `Walk` should work.

5. The `Walk` struct will need to have the RHB and the background, so go ahead and add that:

```
pub struct Walk {
    boy: RedHatBoy,
    background: Image,
}
```

Make sure you've imported `Image` from the `engine` module. Now, you can work your way down the `game` module and follow the compiler errors. In the `initialize` function for `WalkTheDog`, you should see an error for "`expected struct `Walk`, found struct `RedHatBoy`".

6. Fix that by creating `Walk` with the background we already loaded and setting it in `WalkTheDog::Loaded` that's returned. This will look as follows:

```
impl Game for WalkTheDog {
    async fn initialize(&mut self) -> Result<Box<dyn
        Game>> {
        ...
        Ok(Box::new(WalkTheDog::Loaded(Walk {
            boy: rhb,
            background: Image::new(background, Point {
                x: 0, y: 0 }),
        })))
    }
    ...
}
```

This will create `Walk` with a boy and `background` positioned at the upper-left corner, but you should still have several compiler errors in the `update` method of `WalkTheDog` because those all assume that `WalkTheDog::Loaded` contains `RedHatBoy`. Each of those can be changed in the exact same way. The first looks like this:

```
impl Game for WalkTheDog {

    ...

    fn update(&mut self, keystate: &KeyState) {
        if let WalkTheDog::Loaded(walk) = self {
            if keystate.is_pressed("ArrowRight") {
                walk.boy.run_right();
            }

    ...
```

The `if let WalkTheDog::Loaded` line is unchanged, except now the variable name is `walk` instead of `rhb`. Then, we call `run_right` on boy but via the `walk` structure. You could argue that we should add methods to `Walk` instead of delegating to boy, but we'll hold off on that for now. After all, `walk.run_right()` doesn't really make sense. After fixing all the similar compiler errors in `update`, you can also fix a similar error in `draw`, like so:

```
impl Game for WalkTheDog {

    ...

    fn draw(&self, renderer: &Renderer) {
        if let WalkTheDog::Loaded(walk) = self {
            walk.boy.draw(renderer);
        }

    ...
```

Having done all that, you'll now be drawing… well, you'll be drawing RHB again.

7. Next, go ahead and draw the background for our game. Drawing the background is a matter of using our new draw function, so let's add that right before the `walk.boy.draw` function call, as shown here:

```
impl Game for WalkTheDog {

    ...

    fn draw(&self, renderer: &Renderer) {

        if let WalkTheDog::Loaded(walk) = self {
```

```
        walk.background.draw(renderer);
        walk.boy.draw(renderer);
}
...
```

After doing that, you should see RHB standing in front of the background, like this:

Figure 5.1 – Standing in the forest

Looking at it, you might wonder, how come RHB is so far to the right if his x coordinate is 0? Hang on to that thought, as we'll deal with it soon. First, let's get a platform onto the screen, using our sprite sheet from *Chapter 2, Drawing Sprites*.

Adding an obstacle

It's great that we have RHB in front of a background, and it looks great, but the scene is still a little empty. What if there was something else in the scene? Something grand, something innovative, something larger than life. Well, the budget for art is low, so how about a stone?

Our new `Image` class means we won't need much code, and you've seen all of it before. To add an obstacle, follow these steps:

1. Start by copying `Stone.png` from `original/freetileset/png/Object/Stone.png` in the assets and into the `static` directory. Now, you can add it to `Walk` in the same way you added `Background`, like so:

```
struct Walk {
    boy: RedHatBoy,
    background: Image,
    stone: Image,
}
```

That will start causing compiler errors again because `Walk` is created without a stone.

2. In `initialize`, go ahead and load the stone, just as you loaded the background, as shown here:

```
impl Game for WalkTheDog {
    async fn initialize(&mut self) -> Result<Box<dyn
    Game>> {
        ...
        match self {
            WalkTheDog::Loading => {
                ...
                let background = engine::load_image
                    ("BG.png").await?;
                let stone = engine::
                    load_image("Stone.png").await?;
                ...
```

3. Then, you need to take the stone that we just loaded and add it to `Walk`. We'll make sure the stone is on the ground by taking the `FLOOR` value (`600`) and subtracting the height of the stone image, which happens to be `54` pixels. If we position the stone at a *y* position of `546`, it should be sitting right on the ground. Here's the update for creating `Walk`:

```
impl Game for WalkTheDog {
    async fn initialize(&mut self) -> Result<Box<dyn
    Game>> {
```

```
    . . . .

        Ok(Box::new(WalkTheDog::Loaded(Walk {
            boy: rhb,
            background: Image::new(background, Point {
                x: 0, y: 0 }),
            stone: Image::new(stone, Point { x: 150,
                y: 546 }),
        })))
```

4. The stone is 150 pixels to the right, so it will be in front of RHB. Finally, draw the stone using the draw method. That addition is as follows:

```
impl Game for WalkTheDog {
    ...
    fn draw(&self, renderer: &Renderer) {

        if let WalkTheDog::Loaded(walk) = self {
            walk.background.draw(renderer);
            walk.boy.draw(renderer);
            walk.stone.draw(renderer);
        }
```

The code change is small, just drawing the stone with the same call to draw that we've used for boy and background. Do that, and you'll have RHB walking toward the stone:

Figure 5.2 – Look out for that stone!

Now, if RHB walks into that stone, he'll go safely behind it, like this:

Figure 5.3 – Easiest game ever

That's not much fun. While we've learned how to add new objects into the game, and drawn them for a more interactive experience, the game doesn't have any challenges yet. We want the boy to crash into the stone and fall over, ending the game. To do that, we'll need to learn a little about bounding boxes and collision detection, so let's do that in the next section.

Axis-aligned bounding boxes

Checking whether two objects in our game have collided can, theoretically, be done by checking every pixel in every object and seeing whether they share a location. That logic, in addition to being very complicated to write, would be computationally extremely expensive. We need to run at 60 frames a second and can't spend our precious processing power trying to get that kind of perfection – not if we want the game to be fun, anyway. Fortunately, we can use a simplification that will be close enough to fool our silly eyes, the same way we can't tell that animation is really just a series of still images. That simplification is called the *bounding box*.

A bounding box is just a rectangle we'll use for collisions, instead of checking each pixel on the sprite. You can think of every sprite having a box around it, which looks like this:

Figure 5.4 – Bounding boxes

These boxes aren't actually drawn; they only exist in the memory of the game, except when you want to debug them. When you use boxes, you only have to check for the values of the box – top (y), left (x), right (x + width), and bottom (y + height). It makes for a much faster comparison. Let's talk in a little more detail about how to detect when two boxes intersect.

> **Note**
>
> The term "**axis-aligned**" sounds pretty fancy, but all it means is that the boxes aren't rotated. Y will be up and down, X left to right, and always aligned with the game's coordinate system.

Collision

In order to detect whether two boxes collide or overlap, they will exist in the same 2D coordinate space we've been using since the beginning of this book. They may not be visible, but they are there, sitting where the stone is or running along with RHB. They'll need a position in x and y, just like a sprite already has, and also a width and height. When we check whether two boxes are colliding, we check in both the x and y axes. Let's first look at how you can tell whether two boxes intersect in the x axis. Given there are two boxes, box 1 intersects box 2 if the left side (or x position) of box 1 is less than the right side of box 2 but the right side of box 1 is greater than the left side of box 2. This is easier to explain visually:

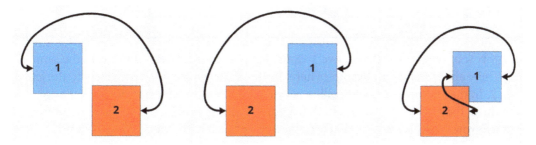

Figure 5.5 – Collisions

The preceding figure shows three sets of two boxes that could potentially collide, in a space where x increases as you move to the right, just like our canvas. The first two comparisons don't collide, but the third does.

Take a look at the first comparison where box 1 is to the left of box 2, with a gap in between them. As you can see, the left side of box 1 is well to the left of the right side of box 2, as shown by the arrows. This passes the first condition for collision – the left side of box 1 must be less than the right side of box 2. However, the right side of box 1 is to the left of box 2's left side, which violates our second condition. To collide, the right side of box 1 must be greater than (to the right) of the left side of box 2, so these two boxes don't collide.

In the second comparison, box 1 has been moved to the right of box 2, again without overlapping. Box 1's right side is now to the right of box 2's left side, so they meet the second condition of colliding, but the left side of box 1 is now also to the right of box 2's right side, so the boxes don't meet the first condition and still don't collide.

Finally, in the third comparison, the left side of box 1 is again to the right of box 2's right side, but the left side of box 1 is to the left of box 2's right side. These two boxes collide. Box 1 and box 2 have overlapping x values, so they collide.

If images aren't your style, it can also help to look at real numbers to see how this algorithm works. Assuming box 1 and box 2 are both 10 x 10, squares we can form a table, like this:

Box 1 left	Box 1 right	Box 2 left	Box 2 right	Intersects?
10	20	21	31	No
11	21	21	31	Yes
31	41	21	31	Yes
32	42	21	31	No

In every row of this table – that is, every example set of coordinates – box 2 is in the same place. There are actually four examples here. In the first row, box 1 is completely to the left of box 2. In the second, the boxes collide because box 1's right edge hits box 2's left edge. In the third, they collide because box 1's left edge is hitting box 2's right edge. Finally, in the fourth row, box 1 is not completely to the right of box 2. The values hold the same properties of the images; either the left edge or right edge of the first box is between the left and right edge of the second box. This long explanation leads to the following short pseudocode:

```
if (box_one.x < box_two.right) &&
    (box_one.right > box_two.x) {
        log!("Collision!");
    }
```

This satisfies two of the conditions I mentioned at the beginning, but what about the vertical axis (y)? That works in a similar way, only instead of using the left and right sides, we use the top and bottom values respectively. The top of box 1 must be above, which means less than, the bottom of box 2. The bottom of box 1 must be below the top of box 2. If both of those are true, the boxes collide. Remember that y goes up as we go down the screen in our coordinate space:

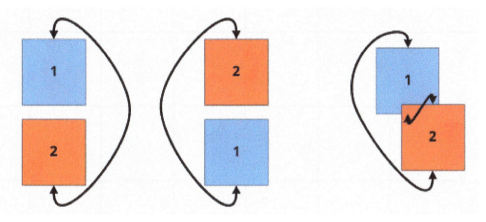

Figure 5.6 – Vertical collisions

Let's take a moment to work through these three comparisons, just as we did before. For the first comparison, the top of box 1 is above the bottom of box 2, but the bottom of box 1 is also above the top of box 2, so they do not overlap.

In the second case, box 1 is completely below box 2, with no collision. The bottom of box 1 is below the top of box 2, which must be true for a collision, but the top of box 1 is also below the bottom of box 2, so our first rule of vertical collisions does not hold.

In the third comparison, the top of box 1 is above the bottom of box 2, and the bottom of box 1 is below the top of box 2, so we have a collision. This means we can extend our pseudocode to look like the following:

```
if (box_one.x < box_two.right) &&
    (box_one.right > box_two.x) &&
    (box_one.y < box_two.bottom) &&
    (box_one.bottom > box_two.y) {
        log!("Collision!");
    }
```

Those are the four things that must be true to get a collision. So, now that we know our collisions, we can apply bounding boxes to RHB and a stone so that they can collide. Unfortunately, a naive approach will lead to really difficult collisions and a nearly impossible game. That problem can be summarized in one word – transparency.

Transparency

In *Figure 5.7*, I've drawn bounding boxes in red for both RHB and the stone:

Figure 5.7 – Bounding boxes

These bounding boxes were created by using the size of the entire sprite after it was loaded, using the width and height properties of HTMLImageElement. As you can see, the boxes are far larger than their corresponding sprites, especially the one for RHB. This is because the sprite has transparency, which we do not want to include in our bounding boxes. Right now, the boxes collide, and RHB would be knocked over by the stone well before touching it. That's not what we want!

This is an example of the primary debugging technique for bounding box collisions – drawing the boxes so that you can see what's wrong. In this case, RHB's box is just way too big. It should be the minimum size required to contain the entire image, and the bug this is revealing is that the sprite sheet we used in *Chapter 2, Drawing Sprites*, contains a lot of transparency. We'll need to fix that before RHB will properly collide with the stone, so let's start trimming the sprite sheet.

Trimming the sprite sheet

In order to have RHB crash into a stone, we're going to have to deal with the transparency. Let's take a look at the raw .png file that RHB is coming from. A portion of the image is shown in *Figure 5.8*, as follows:

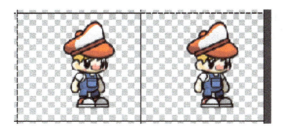

Figure 5.8 – The sprite sheet

This is two frames of the *idle* animation, with black lines showing the image borders. As you can see, there is a **ton** of extra space in these images, so using a bounding box that's the same size as the image won't work. That's the problem you see with the bounding boxes in *Figure 5.7*. We have two choices to fix it. The simplest, although annoying, would be to open our sprite sheet in a graphics editor and find out the actual pixels for the bounding boxes for each sprite. Then, we would store that in code or a separate file and use those bounding boxes. That's faster in development time, but it means loading a much larger image than is necessary and rendering a bunch of transparency for no reason. It's a big performance hit to avoid writing some code, but we might do that if we were in a game jam and needed to finish the game in a hurry.

What we're going to do is use a *trimmed* sprite sheet, which has the transparency taken out. This will mean writing a little code to make sure the sprites still line up, but the memory savings alone (because of a smaller graphic file) will be worth it.

Our trimmed sprite sheet will look like the following (this is a segment):

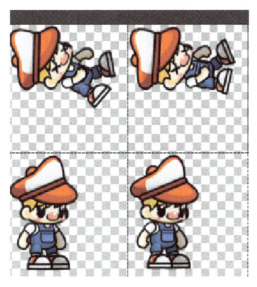

Figure 5.9 – The trimmed sheet

Note that while the white space is trimmed, it's not all removed. That's because each rectangle is still the same size across the entire sheet. Look at how the knocked-out version of RHB takes up the entire rectangle horizontally but the idle RHB takes it up vertically. This means that we'll have to account for some transparency with our bounding box, but fortunately, our sprite sheet JSON will also have that data. We'll also need to make sure that the sprites are lined up properly so that the animations don't jerk around the screen. Fortunately, the JSON provides that data as well.

Note

All the sprite sheets used here are generated with a tool called **TexturePacker**. That includes the JSON that goes along with the graphics. While you can make your own texture maps, why would you? TexturePacker (with both free and paid-for versions) can be found here: `https://bit.ly/3hvZtDQ`. TexturePacker has built-in tools for trimming a sprite sheet and exporting the data we need to make them useful in our game.

The trimmed version of the sprite sheet data file will have a little more information to go along with what we used in *Chapter 2, Drawing Sprites*. Here's an example of the first two idle sprites from the new JSON file:

```
"Idle (1).png":
{
    "frame": {"x":117,"y":122,"w":71,"h":115},
    "rotated": false,
    "trimmed": true,
    "spriteSourceSize": {"x":58,"y":8,"w":71,"h":115},
    "sourceSize": {"w":160,"h":136}
},
"Idle (2).png":
{
    "frame": {"x":234,"y":122,"w":71,"h":115},
    "rotated": false,
    "trimmed": true,
    "spriteSourceSize": {"x":58,"y":8,"w":71,"h":115},
    "sourceSize": {"w":160,"h":136}
},
```

Both frames have the `frame` data that we used previously to cut out our sprite, but they also include a `spriteSourceSize` field. That field contains the bounding box for the non-transparent portion of the sprite. In other words, the first two idle frames had their sprite start with `57` transparent pixels on the left and `8` on top. This information is vital to line up the trimmed sprites, which both start at `0`, `0`. Failing to use this will result in an animation that jumps all over the page and looks terrible. Fortunately, this is rectified by taking the position of the sprite and adding the `spriteSourceSize` x and y coordinates to it. This will result in the sprite not looking like it's in the right place intuitively – that is, when we position the sprite at `0`, it will show up `58` pixels to the right, but as long as we also account for `spriteSourceSize` when doing collision detection, it won't matter. Once we've accounted for `spriteSourceSize`, our bounding boxes will be tight around our sprite sheet, with minimal transparency started:

Figure 5.10 – The correct bounding boxes

> **Note**
>
> If you want to draw your own bounding boxes for debugging, and I recommend that you do, you can add a `draw_rect` function to `Renderer` and draw the rectangle on the context. The code can be found in the source for *Chapter 5, Collision Detection*, at `https://github. com/PacktPublishing/Rust-Game-Development-with- WebAssembly/tree/chapter_5/`.

With these new, corrected bounding boxes, RHB and the stone don't collide, and jumping over the stone safely is eventually possible. In the next section, we'll start by adding the new trimmed sprite sheet.

Adding the trimmed sheet

In the `sprite_sheets` directory of the `assets` folder, you can find new versions of the sprite sheet named `rhb_trimmed.png` and `rhb_trimmed.json`. Copy those over to `static`, but make sure you rename the files `rhb.png` and `rhb.json` respectively. Start your server if it isn't already running, and you should see RHB bouncing around on screen because the sprites in the sheet aren't lined up correctly anymore. He'll also be hovering a little bit over the ground:

Figure 5.11 – Shaking RHB

Our first priority will be to fix up his animation so that it isn't so jerky. This is why we spent so much time discussing `spriteSourceSize` earlier – so that we can fix his animation. First, we'll add that field to `Cell`, which you may or may not recall is in the `engine` module, as shown in the following code snippet:

```
#[derive(Deserialize, Clone)]
#[serde(rename_all = "camelCase")]
pub struct Cell {
    pub frame: SheetRect,
    pub sprite_source_size: SheetRect,
}
```

The changes are the additions of the `#[serde(rename_all)]` directive and a `sprite_source_size` field. While `spriteSourceSize` is the name in the JSON, this is Rust, and in Rust, we use snake case for variable names, which is why we use the `serde(rename_all)` directive. `rename_all = "camelCase"` may seem backward because we're actually renaming to snake case, but that's because the directive refers to serialization, not deserialization. If we were to write out this structure to a JSON file, we'd want to rename any variables to be camelCase, which means to deserialize, we do the opposite. Thanks to the work we did earlier, `sprite_source_size` will be loaded up from the new JSON file, so next, we'll need to adjust the drawing so that the animation lines up again.

In the game module and the RedHatBoy implementation, we'll change the draw function slightly to account for the trimming. It looks like the following:

```rust
impl RedHatBoy {
    ...
    fn draw(&self, renderer: &Renderer) {
        ...
        renderer.draw_image(
        &self.image,
        &Rect {
                x: sprite.frame.x.into(),
                y: sprite.frame.y.into(),
                width: sprite.frame.w.into(),
                height: sprite.frame.h.into(),
            },
        &Rect {
                x: (self.state_machine.context().position.x
                    + sprite.sprite_source_size.x as i16)
                    .into(),
                y: (self.state_machine.context().position.y
                    + sprite.sprite_source_size.y as i16)
                    .into(),
                width: sprite.frame.w.into(),
                height: sprite.frame.h.into(),
            },
        );
    }
```

I've reproduced the entire `draw_image` call for context, but only two lines have changed. Remember that the `draw_image` call takes two rectangles – the source, which is unchanged, and the destination, which is what we changed. Both the *x* and *y* coordinates are adjusted by `sprite_source_size` and its respective coordinates. The cast to `i16` might make you nervous because it could cause math errors if the *x* or *y* position in the sprite sheet is over 2^{15}, but that would be a very strange sheet. Finally, the `into` call is on the computed result, to turn `i16` back into `f32` for the Rect struct. After making those changes, you should see the animation play correctly, and RHB should return to where he was originally, next to the stone:

Figure 5.12 – Good bounding boxes

If you're drawing the bounding boxes with `draw_rect`, make sure it's using the same bounding box as the images. Note how the bounding boxes don't overlap anymore. Still, it's very close, and RHB does hover over the ground a little bit. So, let's adjust his starting position just a bit. At the top of the `red_hat_boy_states` module, we are going to change one constant and add a new one, as follows:

```
const FLOOR: i16 = 479;
const STARTING_POINT: i16 = -20;
```

Previously, FLOOR was `475`, but let's push RHB down just a few pixels. We'll also give RHB a negative *x* position, to give a little room between him and the stone. Remember that RHB is adjusted back to the right to account for animation, so he won't actually be drawn off screen. Next, we'll modify the `RedHatBoyState<Idle>` implementation, specifically the `new` function, to move RHB's starting point. That change is shown here:

```
impl RedHatBoyState<Idle> {
    fn new() -> Self {
        RedHatBoyState {
            context: RedHatBoyContext {
                frame: 0,
                position: Point {
```

```
                    x: STARTING_POINT,
                    y: FLOOR,
                },
                velocity: Point { x: 0, y: 0 },
            },
            _state: Idle {},
        }
    }
}
```

Again, I've included the entire `impl` for context, but the only changes are the initial position of RHB's `RedHatBoyContext`, using the new constants. Do this, and you'll have RHB standing with a little bit of runway so that he can jump the stone, like so:

Figure 5.13 – Get a running start

The bounding boxes are correct in our images, but we're not actually using them yet. That's why if you push the right arrow, RHB will still start running and pass right behind the stone. It's time to give the stone and RHB proper axis-aligned bounding boxes, rather than just drawing them, and then use them to knock RHB right over. What fun!

Colliding with an obstacle

To have collisions, we'll have to actually put the bounding boxes we've seen on both RHB and the stone. Then, in the `update` function of `WalkTheDog`, we'll need to detect that collision, and when that collision happens, we'll move RHB into the `Falling` and `KnockedOut` states, which correspond to the `Dead` animation in the sprite sheet. Much of that code, particularly the state machine, will be very familiar, so I'll refrain from reproducing the parts that are repetitive and highlight the differences. I will remind you of what needs to change in new states, and you can always check the final code at https://github.com/PacktPublishing/Rust-Game-Development-with-WebAssembly/tree/chapter_5/.

Let's start with the easiest bounding box, the one for the stone.

A bounding box for a stone

The stone is the simplest of the bounding boxes because we can just use the size of `HTMLImageElement`. This won't always be the case. If you look at the images of the stone with a bounding box around it, you will notice that it is larger than the stone's actual size, particularly at the corners. For the time being, this will be good enough, but as we proceed, we'll need to keep this in mind.

To add a bounding box to the `Image` implementation, which is in the `engine` module, we'll want to calculate the bounding box when `Image` is created, in its `new` function, as shown here:

```
pub struct Image {
    element: HtmlImageElement,
    position: Point,
    bounding_box: Rect,
}

impl Image {
    pub fn new(element: HtmlImageElement, position: Point) ->
Self {
        let bounding_box = Rect {
            x: position.x.into(),
            y: position.y.into(),
            width: element.width() as f32,
            height: element.height() as f32,
        };
        Self {
            element,
            position,
            bounding_box,
        }
    }

    ....
}
```

Here, we've added `bounding_box` to the `Image` struct, and we construct it in the new function using `width` and `height` from its `HTMLImageElement` backing. It's worth noting that we had to cast the `element.width()` and `element.height()` calls to `f32`. This should be safe, but if later we're drawing a very large image, then it may become a problem. It's also worth noting that by creating the bounding box in the `new` function, we're ensuring that anytime `position` is updated, we also need to update `bounding_box`. We could work around this by calculating `bounding_box` every time, and that's a fine solution, but it does mean potentially losing performance. In this case, we'll keep both `position` and `bounding_box` private in `struct` to ensure they don't get out of sync. `Image` objects don't move yet, anyway.

Given that `bounding_box` is private, we'll need to give it an accessor, so let's do that now:

```
impl Image {
    ...

    pub fn bounding_box(&self) ->&Rect {
        &self.bounding_box
    }
}
```

That takes care of the stone; now, let's give RHB a bounding box.

A bounding box for RedHatBoy

The bounding box on `RedHatBoy` is a little more complicated for the same reasons that the sprite sheet was more complicated. It needs to align with where the sheet is, and it needs to adjust based on the animation. Therefore, we won't be able to do what we did for `Image` and store one `bounding_box` tied to the object. Instead, we'll calculate its bounding box based on its current state and the sprite sheet. The code will actually look very similar to `draw`, as seen here:

```
impl RedHatBoy {
    ...
    fn bounding_box(&self) ->Rect {
        let frame_name = format!(
            "{} ({}).png",
            self.state_machine.frame_name(),
            (self.state_machine.context().frame / 3) + 1
```

```
        );

        let sprite = self
            .sprite_sheet
            .frames
            .get(&frame_name)
            .expect("Cell not found");

        Rect {
            x: (self.state_machine.context().position.x +
                sprite.sprite_source_size.x as i16).into(),
            y: (self.state_machine.context().position.y +
                sprite.sprite_source_size.y as i16).into(),
            width: sprite.frame.w.into(),
            height: sprite.frame.h.into(),
        }
    }
    ...
}
```

To calculate bounding_box, we start by creating frame_name from the state name and the current frame, just like how we did in the draw, and then we calculate Rect from those values using the same calculations we did when we updated the draw function. In fact, it's a good time to clean up some of the duplications in those two pieces of code, using refactoring. Let's extract functions to get the frame and sprite name, still in the RedHatBoy implementation:

```
impl RedHatBoy {
    ...
    fn frame_name(&self) -> String {
        format!(
            "{} ({}).png",
            self.state_machine.frame_name(),
            (self.state_machine.context().frame / 3) + 1
        )
    }
```

```
        fn current_sprite(&self) -> Option<&Cell> {
            self.sprite_sheet.frames.get(&self.frame_name())
        }

        ...
    }
```

For current_sprite, you'll need to make sure you import engine::Cell. Now, we can replace the duplicated code in the bounding_box implementation, as follows:

```
impl RedHatBoy {
    ...

    fn bounding_box(&self) ->Rect {
        let sprite = self.current_sprite().expect("Cell not
            found");

        Rect {
            x: (self.state_machine.context().position.x +
                sprite.sprite_source_size.x as i16).into(),
            y: (self.state_machine.context().position.y +
                sprite.sprite_source_size.y as i16).into(),
            width: sprite.frame.w.into(),
            height: sprite.frame.h.into(),
        }
    }
    ...
}
```

Going further, we can shrink draw by removing the duplicated code from bounding_box and making a much smaller draw function:

```
impl RedHatBoy {
    ...
    fn draw(&self, renderer: &Renderer) {
        let sprite = self.current_sprite().expect("Cell not
            found");

        renderer.draw_image(
            &self.image,
```

```
        &Rect {
            x: sprite.frame.x.into(),
            y: sprite.frame.y.into(),
            width: sprite.frame.w.into(),
            height: sprite.frame.h.into(),
        },
        &self.bounding_box(),
    );
}

...
}
```

This makes for must smaller, cleaner implementations, but it's worth paying attention to the fact that we're looking up current_sprite twice on every frame. We won't work to fix it now because we're not seeing any troubles, but we may want to memoize this value later.

Now that we have both bounding boxes, we can actually see whether RHB collides with the stone.

Crashing on the collision

To crash on a collision, we'll need to check whether the two rectangles intersect using the pseudocode from earlier, only with real code. We'll add that code to Rect, which if you recall is part of the engine module. That code is the implementation on the Rect struct, shown here:

```
impl Rect {
    pub fn intersects(&self, rect: &Rect) -> bool {
        self.x < (rect.x + rect.width)
        && self.x + self.width > rect.x
        && self.y < (rect.y + rect.height)
        && self.y + self.height > rect.y
    }
}
```

This reproduces the previous pseudocode, checking to see whether there is any overlap and returning `true` if there is. Every time you see `rect.x + rect.width`, that's the right side, and `rect.y + height` is the bottom. Personally, I prefer to put the same rectangle on the left-hand side of this function for every condition, as I find it easier to read and think about. We'll use this code in the `update` function of `WalkTheDog`. That code is small, but it will cause a chain reaction. The collision code is as follows:

```
impl WalkTheDog {
    ...
    fn update(&mut self, keystate: &KeyState) {
        if let WalkTheDog::Loaded(walk) = self {
            ...
            walk.boy.update();
            if walk
                .boy
                .bounding_box()
                .intersects(walk.stone.bounding_box())
            {
                walk.boy.knock_out();
            }

        }
    }
}
```

The check for collisions will happen right after the call to `update` on `boy`. We check whether the boy's bounding box has intersected the stone's with our brand new `intersects` function, and if it has, we use `knock_out` on the RHB. Poor RHB; fortunately, you can always refresh.

The `knock_out` function doesn't exist yet; creating it will mean updating our state machine. The `KnockOut` event will cause a transition into the `Falling` state, which will then transition into the `KnockedOut` state when the `Falling` animation has completed. What are we waiting for? Let's knock out RHB!

A KnockOut event

As we did in *Chapter 4*, *Managing Animations with State Machines*, we'll add new states to `RedHatBoyStateMachine` and "follow the compiler" to know where to fill in the necessary code. Rust's type system does a great job of making this kind of work easy, giving useful error messages along the way, so I'm only going to highlight passages that are unique. Remember that you can always peek ahead using the source code at `https://github.com/PacktPublishing/Rust-Game-Development-with-WebAssembly`, although I highly recommend you try writing the implementation yourself first.

You can get started in the game module by adding a `KnockOut` event to `Event` enum and a `knock_out` method onto `RedHatBoy` as with the other state machine transitions, as shown below:

```
pub enum Event {
    Run,
    Jump,
    Slide,
    KnockOut,
    Update,
}
...
impl RedHatBoy {
    ...
    fn knock_out(&mut self) {
        self.state_machine =
            self.state_machine.transition(Event::KnockOut);
    }
    ...
```

This will just move the compiler error into `RedHatBoyStateMachine` because match statements are incomplete, so you'll need to add a `KnockOut` event to `RedHatBoyStateMachine` that will transition from `Running` to `Falling`. That transition is like so:

```
impl RedHatBoyStateMachine {
    fn transition(self, event: Event) -> Self {
        match (self, event) {
            ...
```

```
            (RedHatBoyStateMachine::Running(state),
                Event::KnockOut) => state.knock_out
                    ().into(),
            (RedHatBoyStateMachine::Jumping(state),
                Event::KnockOut) =>
                    state.knock_out().into(),
            (RedHatBoyStateMachine::Sliding(state),
                Event::KnockOut) =>
                    state.knock_out().into(),
            _ => self,
        }
    }
}
...
```

You might wonder why we also have transitions from Jumping and Sliding to Falling; that's because if we don't do that, then the user can simply hold down the spacebar to jump continuously, or slide at the right time, and they will pass right through the stone. So, we need to make sure that all three of those states will transition to Falling in order for the game not to have any bugs.

Of course, there's still a lot missing. Falling doesn't exist yet, neither as a member of the RedHatBoyStateMachine enum nor as a struct. The typestates for Sliding, Jumping, or Running don't have knock_out methods, and there's no From trait implemented to convert from Falling into RedHatBoyStateMachine::Falling. You'll need to add both of those, just like before, and fill in the rest of the compiler errors. You'll find that you need two new constants, the number of frames in the falling animation and the name of the falling animation in the sprite sheet. You can look at rhb.json and figure out the values, or look at the following listings:

```
const FALLING_FRAMES: u8 = 29; // 10 'Dead' frames in the
sheet, * 3 - 1.
const FALLING_FRAME_NAME: &str = "Dead";
```

If you've made all the proper boilerplate changes, you'll end up making a transition from Running to Falling that looks like the following code:

```
impl RedHatBoyState<Running> {
    pub fn knock_out(self) -> RedHatBoyState<Falling> {
        RedHatBoyState {
            context: self.context,
```

```
                        _state: Falling {},
            }
        }
    ...
```

Note that you're only transitioning states at this point, not making any changes to
`RedHatBoyContext`. This is why things get weird because when RHB collides with the
stone, he falls over... and keeps sliding and falling over forever:

Figure 5.14 – Sliding while falling?

The transition properly moves into the `Dead` animation, but it doesn't stop RHB's forward
motion. Let's change the transition to stop `RedHatBoy`:

```
impl RedHatBoyState<Running> {
    pub fn knock_out(self) -> RedHatBoyState<Falling> {
        RedHatBoyState {
            context: self.context.reset_frame().stop(),
            _state: Falling {},
        }
    }
    ...
```

Now, when setting the new state, we call `reset_frame()` to set the frame to 0, as we
always do when changing animations, and call the new `stop` function that will halt the
character's forward motion. Of course, that function isn't written yet. It's attached to the
`RedHatBoyContext` implementation, setting the `velocity.x` to 0:

```
impl RedHatBoyContext {
    fn stop(mut self) -> Self {
        self.velocity.x = 0;
```

```
                    self
            }
        }
    ...
```

You'll want to do the same transition when going from Sliding to Falling and Jumping to Falling as well so that the transitions match. That will halt the character's forward motion but will not stop the death animation from playing over and over again. That's because we never transition out of the Falling state and into KnockedOut, which itself doesn't exist yet. Fortunately, we've done code like this before. Remember in *Chapter 4, Managing Animations with State Machines*, we transitioned out of the Sliding animation and back into the Running animation when the slide animation was complete. That code, which is in the update function of RedHatBoyState<Sliding>, is reproduced here:

```
impl RedHatBoyState<Sliding> {
    ...
    pub fn update(mut self) -> SlidingEndState {
        self.update_context(SLIDING_FRAMES);
        if self.context.frame >= SLIDING_FRAMES {
            SlidingEndState::Running(self.stand())
        } else {
            SlidingEndState::Sliding(self)
        }
    }
}
```

In this code, we check every update and see whether the Sliding animation is complete via if state_machine.context.frame>= SLIDING_FRAMES. If it is, we return the Running state instead of the Sliding state. In order to get this far, you already had to add an update method to RedHatBoyState<Falling>, likely with a generic default that played the animation. Now, you'll need to mimic this behavior and transition into the new KnockedOut state. Specifically, you'll need to do the following:

1. Create a KnockedOut state.
2. Create a transition from Falling to KnockedOut.
3. Check in the update action whether the Falling animation is complete, and if so, transition to the KnockedOut state instead of staying in Falling.

4. Create an `enum` to handle both end states of the `update` method in
    `RedHatBoyState<Falling>`, as well as the corresponding `From` trait, to
    convert from that to the `RedHatBoyStateMachine`-appropriate `enum` variant.

The only thing new here is that `RedHatBoyState<KnockedOut>` will not need
the `update` method because, in the `KnockedOut` state, RHB doesn't do anything.
We won't go through that code step by step, and instead, I highly encourage you to
try it yourself. If you get stuck, you can look at the code at `https://github.com/
PacktPublishing/Game-Development-with-Rust-and-WebAssembly/
tree/chapter_5`. When you're done, it should look like this:

Figure 5.15 – Just taking a nap

In the meantime, I'll assume you did it because you're awesome, so we'll move on to
jumping onto a platform.

Jumping onto a platform

Now that RHB crashes into a stone, we'll need to find a way to go over it. Play the game
and try jumping the rock; you'll that notice it's really difficult. The timing has to be just
right, reminiscent of the scorpions in the classic game *Pitfall* for the Atari 2600. Later
in this chapter, we'll adjust that by shrinking the bounding boxes and increasing the
horizontal speed of RHB, but first, we're going to put a platform above the stone that RHB
can jump on to avoid the rock. In addition to putting a platform on screen with a new
sprite sheet and giving it a bounding box, we'll have to handle a new type of collision.
Specifically, we'll need to handle collisions coming from above the platform so that we can
land on it.

Adding a platform

We'll start by adding the platform from a new sprite sheet. This sprite sheet actually contains the elements that will make up our map in the upcoming chapters, but we'll use it for just one platform for now. The sprite sheet looks like this:

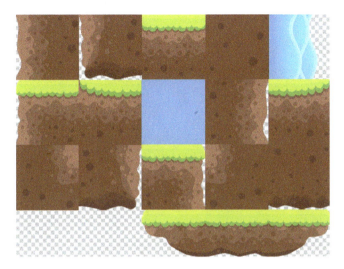

Figure 5.16 – Our platforms

The image is divided up into squares that aren't outlined but are visible in the way the shapes are arranged, called tiles. Those squares are the sprites that we'll be mixing and matching to make various obstacles for RHB to jump over and slide under. The tiles are also jammed together nice and tight, so we won't have to concern ourselves with any offsets. For the time being, we'll only need the platform at the lower-right corner, which will float over the stone:

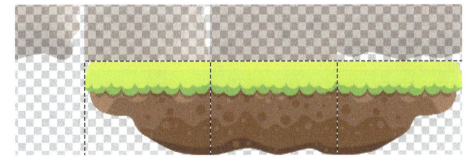

Figure 5.17 – One platform

This one is conveniently set up with the sprites in order, so it will be easy to access in the sprite sheet. You can see those dotted lines now marking the three sprites. Let's get it into our game. In the `sprite_sheets` directory of the assets, you'll find two files, `tiles.json` and `tiles.png`. This is the sheet for the tiles, which we'll need to load at startup. So that we have something to load it into, we'll start by creating a `Platform` struct in the game module:

```
struct Platform {
    sheet: Sheet,
    image: HtmlImageElement,
    position: Point,
}

impl Platform {
    fn new(sheet: Sheet, image: HtmlImageElement, position:
        Point) -> Self {
        Platform {
            sheet,
            image,
            position,
        }
    }
}
```

So far, this just loads up the expected data. At this point, you may note that `sheet` and `image` are paired together repeatedly, which means they are good candidates for refactoring into a new structure, such as `SpriteSheet`. We won't do that now because we don't want to be premature and refactor to a bad abstraction, but we'll keep an eye out for the duplication if it shows up again.

The platform is going to need two things. It's going to need to be drawn, and it's going to need a bounding box so that we can land on it. To draw the box, we'll need to draw the three tiles that make that platform on the bottom together. Looking at `tiles.json`, it's hard to tell which platforms we want because the frame names are all just numbers such as `14.png`, so just take my word for it that the tiles are `13.png`, `14.png`, and `15.png`.

> **Note**
>
> It's significantly easier to tell which tiles you need to look at using a tool such as TexturePacker. If you don't have that available, you can just draw each of the images from the sheet, with their names displayed as well, and then modify their names in the JSON file to be more readable.

Let's dive into the `draw` function for `Platform` now, which has a little trick in it, as seen here:

```rust
impl Platform {
    ...
fn draw(&self, renderer: &Renderer) {
        let platform = self
            .sheet
            .frames
            .get("13.png")
            .expect("13.png does not exist");

        renderer.draw_image(
        &self.image,
        &Rect {
                x: platform.frame.x.into(),
                y: platform.frame.y.into(),
                width: (platform.frame.w * 3).into(),
                height: platform.frame.h.into(),
            },
            &Rect {
                x: self.position.x.into(),
                y: self.position.y.into(),
                width: (platform.frame.w * 3).into(),
                height: platform.frame.h.into(),
            },
        );
}
}
```

The cheat is that we know that the three tiles happen to be next to each other in the sheet, so instead of getting all three sprites out of the sheet, we'll just get three times the width of the first sprite. That will happen to include the other two tiles. Don't forget that the second `Rect` is the destination and, as such, should use the `position` field. That second rectangle also corresponds to the bounding box of the platform, so let's create the platform's bounding box function and use it there instead. These changes are shown here:

```
impl Platform {
    ...

    fn bounding_box(&self) ->Rect {
        let platform = self
            .sheet
            .frames
            .get("13.png")
            .expect("13.png does not exist");

        Rect {
            x: self.position.x.into(),
            y: self.position.y.into(),
            width: (platform.frame.w * 3).into(),
            height: platform.frame.h.into(),
        }
    }

    fn draw(&self, renderer: &Renderer) {
        ...
        renderer.draw_image(
            &self.image,
            &Rect {
                x: platform.frame.x.into(),
                y: platform.frame.y.into(),
                width: (platform.frame.w * 3).into(),
                height: platform.frame.h.into(),
            },
            &self.bounding_box(),
```

```
        );
    }
```

This code has the same troubles as other code where we search for the frame on every draw and we're doing it twice. We're also constructing `Rect` on every `bounding_box` call, which we explicitly avoided earlier. Why the change? Because I know the future, and we'll be changing how we construct this shortly, so it's not worth worrying about saving an extra cycle or two here. Trust me.

Now that we've made a platform that could theoretically be drawn, let's actually draw it. First, we'll add it to the `Walk` struct, as shown here:

```
struct Walk {
    boy: RedHatBoy,
    background: Image,
    stone: Image,
    platform: Platform,
}
```

Of course, that won't compile because when we create `Walk`, we don't have a platform. We need to update the `initialize` function in `WalkTheDog` to include the new `Platform`, as shown here:

```
impl Game for WalkTheDog {
    async fn initialize(&mut self) -> Result<Box<dyn Game>> {
        match self {
            WalkTheDog::Loading => {
                ...
                let stone = engine::
                    load_image("Stone.png").await?;

                let platform_sheet = browser::
                    fetch_json("tiles.json").await?;

                let platform = Platform::new(
                    platform_sheet.into_serde::<Sheet>()?,
                    engine::load_image("tiles.png").await?,
                    Point { x: 200, y: 400 },
                );
```

```
    . . .
        Ok(Box::new(WalkTheDog::Loaded(Walk {
            boy: rhb,
            background: Image::new(background,
                Point { x: 0, y: 0 }),
            stone: Image::new(stone, Point { x:
                150, y: 546 }),
            platform,
        })))
    . . .
```

There are only a few small changes here, which I've highlighted. We then fetch the
tiles.json and create a new Platform with it and tiles.png. Finally, we create
Walk with platform. Drawing the platform is a one-line change, adding it to the draw
function of WalkTheDog, as shown here:

```
fndraw(&self, renderer: &Renderer) {
    . . .

    if let WalkTheDog::Loaded(walk) = self {
        walk.background.draw(renderer);
        walk.boy.draw(renderer);
        walk.stone.draw(renderer);
        walk.platform.draw(renderer);
    }
}
```

If you've done this correctly, you should see the following:

Figure 5.18 – An escape!

But while the platform has a bounding box, you aren't using it yet, so we'll need to add that collision to the update function of WalkTheDog. When colliding with the platform, you'll want to transition from Jumping back to Running. This transition is already written – we do it when we land on the floor – so you'll just need to add a check and an event that can perform the transition.

We'll also need to make sure that RHB stays on the platform. Currently, gravity would just pull him right through it, regardless of whether or not there's a collision or the player is in the Running state. That solution is a little more complex. A naive solution, and I know because I wrote it, is to stop applying gravity when the player is on the platform. This works until it doesn't, causing a **Wile E. Coyote** effect when RHB runs off the platform and stays in the air. Presumably, if he could look down, he would hold up a sign and then crash to the ground.

Instead, what we do is continue to apply gravity on every frame and check whether RHB is still landing on the platform. If he is, then we adjust him right back onto the top of it. This effectively means that RHB "lands" repeatedly until he reaches the end of the platform, when he falls off. Fortunately, this isn't visible to the user, since we calculate RHB's new position on every update, and this results in him moving to the right until he falls off the edge, as he should.

Let's start by adding the check to the update function so that RHB can land on a platform:

```
fn update(&mut self, keystate: &KeyState) {
    if let WalkTheDog::Loaded(walk) = self {
        ...
        walk.boy.update();

        if walk
            .boy
            .bounding_box()
            .intersects(&walk.platform.bounding_box())
        {
            walk.boy.land();
        }

        if walk
            .boy
            .bounding_box()
```

```
            .intersects(walk.stone.bounding_box())
        {
            walk.boy.knock_out()
        }
        ...
    }
}
```

I've reproduced the check for the boy intersecting the stone as well so that you can see that we checked the bounding box before checking the stone. It doesn't really matter which check comes first, but I prefer to check things that can kill the player last. That way we won't kill the player when we really want them to land on a platform. Just as when we created the knock_out method on RedHatBoy, the land method and its corresponding Event don't exist yet. You can create them both now, and follow the compiler until you have to write the transition in the state machines, as shown here:

```
impl RedHatBoyStateMachine {
    fn transition(self, event: Event) -> Self {
        match (self, event) {
            (RedHatBoyStateMachine::Jumping(state),
Event::Land) => {
                state.land().into()
            }
            ...
```

Remember that we already wrote a transition method from Jumping to Running, so you won't need to write it, but as I mentioned previously, this isn't enough to land on the platform. The transition will happen, but RHB will fall right through the platform and crash into the ground. Not cool. In order to keep RHB on the platform, we need to set its *y* position to the top of the bounding box. This will mean changing the Land event to store the *y* position of the platform's bounding box.

> **Tip**
> Because we used enum for the events, we can pass any data we need by adding it as part of the variant we are using. The Rust enum is a great feature of Rust.

On every intersection with the platform, we'll transition with the Land event. This means that the Update event will pull the player down a bit because of gravity, but then the Land event will push them right back up. Since we don't draw the in-between state, it will look fine. This system isn't perfect, but we aren't writing a physics engine. Let's do that now; we'll start by modifying the land function to be land_on, taking a *y* position:

```
impl Game for WalkTheDog {
    ...
    fn update(&mut self, keystate: &KeyState) {
        ...
        if walk
            .boy
            .bounding_box()
            .intersects(&walk.platform.bounding_box())
        {
            walk.boy.land_on(walk.platform.bounding_box().y);
        }
    }
    ...
}
```

Now, land_on instead of land takes the *y* position of bounding_box for the platform. If you just follow the compiler errors for that, you will eventually need to modify the Land event to hold the position and modify the land method on the Jumping typestate. It will probably look something like this:

```
impl RedHatBoyState<Jumping> {
    ...
    pub fn land_on(mut self, position: f32) ->
        RedHatBoyState<Running> {
            self.context.position.y = position as i16;
            RedHatBoyState {
                context: self.context.reset_frame(),
                _state: Running,
            }
        }
}
```

As an initial attempt, this seems fine. It's unfortunate that `self` had to be made mutable, but the transition sets RHB's position back to the top of the platform. The problem is that the *y* position of RHB actually represents his top-left corner. This means that if you followed this to its conclusion, you'd get something like this:

Figure 5.19 – This does not look right

Fortunately, `RedHatBoy` knows his height, so we can adjust for the height when setting the *y* position. We would need to include `self.bounding_box.height()` as a parameter in the `Land` event and then account for it during the transition, like so:

```
impl RedHatBoyState<Jumping>{

    ...

    fn land_on(mut self, position: f32, height: f32) {
        let position = (position - height) as i16;
        RedHatBoyState {
            context: self.context.reset_frame(),
            _state: Running,
        }
    }
}
```

This sort of works, but it has another problem. The bounding box is actually changing size during the animation, based on the current frame of the animation, because the trimmed sprite shrinks and grows slightly. As we check collisions on every frame, we'll call `Land` repeatedly while RHB is on the platform. If we continually change the landing position based on the current frame's height, the walk ends up looking very "bouncy." Even though the bounding box is changing slightly, it looks better if we use a constant value for the player's height for this calculation.

> **Tip**
> Game development frequently has a lot of trial and error. When the
> mathematically correct solution doesn't play well or look right, remember that
> the feel of the game is more important than mathematical accuracy.

We already have the player height adjustment; we just created it as the FLOOR constant.
In the game module, you'll see that the FLOOR constant is set at 479. Well, that means
that we can use the height of the game (which is 600) and subtract FLOOR to get the
player's height. We can use that info to create two new constants. The first, HEIGHT, can
be defined in the game module as const HEIGHT: i16 = 600 and used wherever
we've hardcoded the 600 value. The second, PLAYER_HEIGHT, can be defined in the
red_hat_boy_states module, as shown here:

```
mod red_hat_boy_states {
use super::HEIGHT;

    ...

    const FLOOR: i16 = 479;
    const PLAYER_HEIGHT: i16 = HEIGHT - FLOOR;
```

PLAYER_HEIGHT belongs in the red_hat_boy_states module, since it will only be
used there, but to calculate it, we need to import game::HEIGHT into the red_hat_
boy_states module. We do that with the highlighted use statement. Now that we
have the proper value to adjust RHB when he lands, we can account for it in the land_on
method and RedHatBoyContext:

```
impl RedHatBoyContext {
    ...
    fn set_on(mut self, position: i16) -> Self {
        let position = position - PLAYER_HEIGHT;
        self.position.y = position;
        self
    }
}
...
impl RedHatBoyState<Jumping> {
    pub fn land_on(self, position: f32) ->
        RedHatBoyState<Running> {
        RedHatBoyState {
            context: self.context.reset_frame()
```

```
            .set_on(position as i16),
            _state: Running,
        }
    }
...
```

We've moved the adjustment of RHB's position into a `set_on` method in `RedHatBoyContext`. The `set_on` method always adjusts for the player's height, which is why it's named `set_on` and not `set_position_y`. It also returns `self` so that we won't require `mut self` anymore, fitting with the rest of the operations on `RedHatBoyContext`.

Changing the `land` method to the `land_on` method also requires you to modify what it is called within the `update` method of `RedHatBoyState<Jumping>`. After all, there is no `land` method anymore. Keep in mind that we have to account for the height when calling `set_on`, as shown here:

```
impl RedHatBoyState<Jumping> {
    pub fn update(mut self) -> JumpingEndState {
        self.update_context(JUMPING_FRAMES);

        if self.context.position.y >= FLOOR {
            JumpingEndState::Landing(self.land_on
                (HEIGHT.into()))
        } else {
            JumpingEndState::Jumping(self)
        }
    }
}
```

Here, we are checking whether RHB is past `FLOOR` and pushing it back up to `HEIGHT`. Remember that When we call land_on we send the position of RHB's feet, not his head. You could argue that the `update` method shouldn't check for hitting the ground and that the higher-level `update` method in `WalkTheDog` should check for collisions with the ground and use the `Land` event when appropriate. I think I'd agree, but we've made more than enough changes for this chapter, so we'll stick with it as it is for now.

This adjusts the position of RHB for landing. He'll be positioned on the platform or the ground at the end of his jump. Now, we need to make sure that the `Land` event prevents RHB from falling through the platform right after he lands. The `Land` event will happen while `Running` occurs on the platform, but it isn't handled, so you'll fall right through because gravity takes effect. We're going to need a `Land` transition for every state that is valid on the platform, where the state stays the same but the `y` position is forced back to the top of the platform.

> **Note**
>
> If I might steal a line from the *Big Nerd Ranch* series of books, programming is hard and you are not stupid. It may appear that these changes emerged fully formed because I am a super-expert, but in many cases, these only came about through much trial and error, rereading old books, and luck. So, don't worry if you wouldn't have come up with this solution off the top of your head or things have gotten a little confusing. Take another try at the code, slow down, and have fun. We're making a game!

Fortunately, it's harder to explain why we need this code than to actually write it. We'll handle the `Land` event for `Running` in the `transition` method:

```
impl RedHatBoyStateMachine {
    fn transition(self, event: Event) -> Self {
        match (self, event) {
            ...
            (RedHatBoyStateMachine::Running(state), Event::
                Land(position)) => {
                state.land_on(position).into()
            }
```

Then, we'll add a `land_on` method to the `RedHatBoyState<Running>` typestate, as shown here:

```
...
impl RedHatBoyState<Running> {
    ...
    pub fn land_on(self, position: f32) ->
        RedHatBoyState<Running> {
        RedHatBoyState {
            context: self.context.set_on(position as
```

```
                   i16),
               _state: Running {},
        }
     }
  }
```

For every `Land` event in the `Running` state, you adjust the position and stay in the
`Running` state. With that, you should see RHB jump onto a platform:

Figure 5.20 – Running on the platform

Running on the platform is beginning to work, but you'll find a strange bug if you try to
run past the edge of the platform. RHB falls through the bottom!

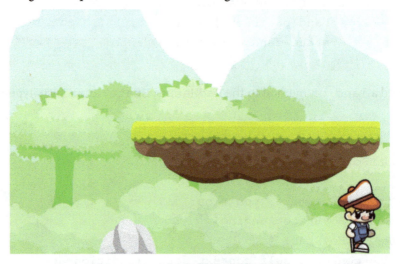

Figure 5.21 – My God! How did I get here?

It turns out there is a fairly sneaky bug with the way we are handling gravity, which we'll call the "terminal velocity" bug, and we can address that next.

Terminal velocity

If you log RHB's `velocity` in *y* in the `update` method as you jump on the platform and walk across it, it looks like this:

```
Gravity 75                      index_bg.js?6119:391
Gravity 76                      index_bg.js?6119:391
Gravity 77                      index_bg.js?6119:391
Gravity 78                      index_bg.js?6119:391
Gravity 79                      index_bg.js?6119:391
Gravity 80                      index_bg.js?6119:391
Gravity 81                      index_bg.js?6119:391
Gravity 82                      index_bg.js?6119:391
Gravity 83                      index_bg.js?6119:391
Gravity 84                      index_bg.js?6119:391
Gravity 85                      index_bg.js?6119:391
Gravity 86                      index_bg.js?6119:391
Gravity 87                      index_bg.js?6119:391
Gravity 88                      index_bg.js?6119:391
Gravity 89                      index_bg.js?6119:391
Gravity 90                      index_bg.js?6119:391
Gravity 91                      index_bg.js?6119:391
Gravity 92                      index_bg.js?6119:391
Gravity 93                      index_bg.js?6119:391
```

Figure 5.22 – Gravity forever!

If you recall, we add 1 to the gravity on every update until the player jumps again. This means that, eventually, the gravity gets so large that the player is pulled completely below the platform on an update, and he actually stops intersecting it. Our platform is currently at 400. When the player lands on it, he is at 279, the platform's *y-axis* minus the player's height. On the first frame, we pull him down by 1 for gravity, check whether that intersects the platform (it does), and land. On the next frame, we pull him down by 2, the next by 3, and so on. Eventually, we actually pull him completely beneath the platform, he does not intersect it, and boom – he's suddenly below the platform. We need to fix that by giving gravity a **terminal velocity**.

In the real-world, terminal velocity is the fastest attainable speed by an object as it falls because of the drag of the air around it (see `https://go.nasa.gov/3roAWGL` for more information). We aren't going to calculate RHB's true terminal velocity, as there's no air in his world, but we can use the very scientific method of picking a number and seeing whether it works. We'll set a maximum positive *y* velocity of RHB to 20 and clamp his updates to that. That will live in the `RedHatBoyContext` `update` method, where we are already modifying *y* for gravity. The code for that is shown here:

```
mod red_hat_boy_states {

    ...

    const TERMINAL_VELOCITY: i16 = 20;

    ...

    impl RedHatBoyContext {
        pub fn update(mut self, frame_count: u8) -> Self {
            if self.velocity.y < TERMINAL_VELOCITY {
                self.velocity.y += GRAVITY;
            }

    ...
```

Clamping the velocity at 20 fixes our issue with falling through the platform, and now RHB falls off the platform at the end as he should. However, if you try to slide (push the arrow down), you'll see that RHB falls right through the platform. That's because the `Sliding` state doesn't respond to the `Land` event. You can fix that in the exact same way you fixed `Running`, which is an exercise for you. Give it a try, and remember that if you get stuck, the final source code is available at `https://github.com/PacktPublishing/Game-Development-with-Rust-and-WebAssembly/tree/chapter_5`. One hint – when you stay in the same state, you don't call `reset_frame`!

That's almost the end of it, but there are two more things to take care of – crashing into the bottom of the platform and transparency in the bounding boxes.

Collision from below

At the moment, if RHB collides with the platform, he is set on the top, which is great for landing but not so great if he comes from beneath the platform. If you were to comment out the collision with the stone right now and run straight ahead, you'd actually find yourself suddenly pop up onto the platform! Why? Because RHB's head actually bumps into the bottom of the platform, and that collision causes the `land_on` event to fire. Instead of banging his head and falling over, he teleports onto the platform!

We need to have special collision detection here. RHB can only land on the platform if he comes from above it; otherwise, it's game over. Fortunately, this can be handled in the `update` function with two small changes to the way we check collisions. Collisions with the platform where `RedHatBoy` is *above* the platform means landing; otherwise, it's the same as hitting a stone, and you get knocked out. You also need to be *descending*; otherwise, you'll get this weird effect where you stick to the platform while still going up in your jump. Let's see that change:

```
impl Game for WalkTheDog {
    ...
    fn update(&mut self, keystate: &KeyState) {
        if let WalkTheDog::Loaded(walk) = self {
            ...
            walk.boy.update();
            if walk
                .boy
                .bounding_box()
                .intersects(&walk.platform.bounding_box())
            {
                if walk.boy.velocity_y() > 0 && walk.boy.
pos_y() < walk.platform.position.y {
                    walk.boy.land_on(walk.platform.bounding_
box().y);
                } else {
                    walk.boy.knock_out();
                }
            }
            ...
```

```
            }
        }
    }
```

The changes are to check whether the RedHatBoy velocity, in *y*, is greater than 0 and, therefore, RHB is moving down. We also check whether the position in *y* is less than the top of the platform's *y* position. This means that the boy is above the platform, so he's landing on it; otherwise, the boy has crashed into it, and we knock him out. The pos_y and velocity_y functions don't exist yet, so we'll add those to RedHatBoy, as shown here:

```
impl RedHatBoy {
    ...
    fn pos_y(&self) -> i16 {
        self.state_machine.context().position.y
    }

    fn velocity_y(&self) -> i16 {
        self.state_machine.context().velocity.y
    }
    ...
}
```

It's a little tricky to get the *y* values for RedHatBoy because they are actually on RedHatBoyContext, but we are able to pull it off here and wrap them in a getter for convenience.

> **Info**
>
> For the sake of the book, the code here is pretty explicit, but you can make it more expressive by extracting a method for falling on RedHatBoy. We'll leave it as it is for now, but you will want to consider some more expressive names in your own code.

With that, RHB can finally run, jump over stones, land on platforms, and fall off them. However, you've probably noticed that the collisions are really crude. He crashes into the bottom of the platform easily because the transparent parts of the images collide. He also can walk past the edge of the platform, again because of the transparent parts of the image:

Figure 5.23 – Believe it or not, I'm walking on air

Let's spend a little time tweaking our bounding boxes to deal with the transparency.

Transparency in bounding boxes

The problem with our bounding boxes is that we're using the image dimensions as the bounding box. That means we'll have a lot of extra space around our characters for our bounding boxes. In this screenshot, I've used the draw_rect method from earlier in this chapter to show the bounding boxes for all three objects in our scene:

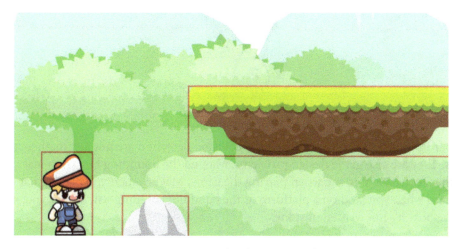

Figure 5.24 – Bounding boxes everywhere

The platform has a lot of white space in the bounding box, particularly at the lower-left and lower-right corners. RHB also has white space near the corners of his hat. When we turn off the collision checks on the stone and try to walk under the platform, RHB "collides" with the lower-left corner of the platform well before he actually hits it.

The white space around RHB's feet is a problem too; they are what cause the landing on air effect. The far-right edge of his bounding box intersects with the platform, so he lands before he's really in the right position. If you could see him walk off the edge of the platform, you'd see that it has the same problem when he walks off. He takes several steps in mid-air before he begins to fall.

We'll start by dealing with RHB's bounding box to make landing and falling off the platform look a little more realistic.

Fixing the game to fit

There are algorithms we can use to make the bounding box better match the actual pixels in the image, but ultimately, none of them are necessary. Spend a little time with most platformers and you'll see that the collisions aren't perfect, and 99% of the time, it's just fine. After spending a little time "researching" by playing video games, I determined that if we simply make the box only as wide as the feet, he develops a much more realistic landing. This is a little counter-intuitive. If we narrow the box, his arm and hat will stick out past the edge of the box; we'll miss collisions! Does this matter?

Figure 5.25 – A narrow bounding box

The answer is, "maybe." Bounding boxes and collision detection are not just mathematical problems. They are also game design problems. Making the bounding box wrap just around the feet felt right to me when playing the game. Maybe when you play it, that will feel too hard when you land on a platform or the hand not colliding will bother you, so change the box! It's not written in stone.

After experimenting, I found that I wanted to shorten the box as well so that RHB couldn't be knocked out by grazing his hat. To mimic that, we can start by renaming `bounding_box` `destination_box`, because that represents where the sprite is rendered *to*. It needs to be at the position of `RedHatBoy` in the game but with the width and height of the source image; otherwise, the image will appear squished. Then, we can re-implement the `RedHatBoy` bounding box, like so:

```
impl RedHatBoy {
    ...
    fn bounding_box(&self) -> Rect {
        const X_OFFSET: f32 = 18.0;
        const Y_OFFSET: f32 = 14.0;
        const WIDTH_OFFSET: f32 = 28.0;
        let mut bounding_box = self.destination_box();
        bounding_box.x += X_OFFSET;
        bounding_box.width -= WIDTH_OFFSET;
        bounding_box.y += Y_OFFSET;
        bounding_box.height -= Y_OFFSET;
        bounding_box
    }
```

We start with the original dimensions of the image, `destination_box`, and simply shrink it by some offsets. I chose the numbers by using the high-tech system of picking numbers and looking at them. This gave me a bounding box that looked natural jumping and falling off the cliff while not being so small that RHB never hits anything.

If you did a global find and replace on `bounding_box` and changed it to `destination_box`, then the collision detection is incorrect. We need to use `bounding_box` for checking collisions and `destination_box` for drawing. Drawing should already be complete; you'll need to go into the `update` method in `WalkTheDog` and make sure that every `intersects` call is on the `bounding_box`, not `destination_box`.

With the new `bounding_box` method and a properly drawn image, you get a bounding box that looks like this for RHB:

Figure 5.26 – A tight-fit bounding box

You can see that it's a lot smaller than the image, which makes the game look better and play a little more forgiving. He lands on and falls off the platform much more accurately, without a hovering effect. You also might find it easier to jump the stone now because the transparent part of RHB doesn't crash into the rock.

That leaves us with the white space around the edges of the platform. We can shrink the bounding box for it, but that would cause the player to fall through the top of the platform when he lands on the edges. The platform is narrower on the bottom than the top, which is a problem because we crash into the bottom and land on the top. What we really want to do is take the platform and make it into multiple bounding boxes.

Subdividing bounding boxes

Subdividing the bounding boxes is just like it sounds – we're going to take the one bounding box that is currently used for the platform and divide it into several. This will dramatically reduce the amount of extra space in the boxes and improve our collision detection. You might think that we'll use a complex algorithm or tool to determine what boxes to use, and we will – it's our eyeballs.

Specifically, we'll look at the platform, see the white space, and then try out a few versions of the bounding boxes divided up until we find a solution we like. We can begin that process by making it possible for `Platform` to have more than one bounding box. We'll do that by, again, renaming `bounding_box` to `destination_box` and then creating a new method to construct a vector of `bounding_boxes` from that original box, as shown here:

```
impl Platform {
    fn bounding_boxes(&self) -> Vec<Rect> {
        const X_OFFSET: f32 = 60.0;
        const END_HEIGHT: f32 = 54.0;
```

```
        let destination_box = self.destination_box();
        let bounding_box_one = Rect {
            x: destination_box.x,
            y: destination_box.y,
            width: X_OFFSET,
            height: END_HEIGHT,
        };

        let bounding_box_two = Rect {
            x: destination_box.x + X_OFFSET,
            y: destination_box.y,
            width: destination_box.width - (X_OFFSET *
                2.0),
            height: destination_box.height,
        };

        let bounding_box_three = Rect {
            x: destination_box.x + destination_box.width -
                X_OFFSET,
            y: destination_box.y,
            width: X_OFFSET,
            height: END_HEIGHT,
        };

        vec![bounding_box_one, bounding_box_two,
            bounding_box_three]
    }
```

In this method, we create three Rects, each meant to match the platform, starting from the destination box of the platform. It's two small rectangles on the edges and one bigger one in the middle. When we draw those boxes, it looks like this:

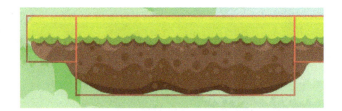

Figure 5.27 – Platform bounding boxes

That's a lot less white space that you can collide with. You might wonder where the values for X_OFFSET and END_HEIGHT came from, and the truth is that I just drew the boxes and looked at them until I was happy. It's not fancy; it's just good enough.

Now that we're using a vector of bounding boxes instead of just one, we'll need to change the logic in the update method of WalkTheDog to make sure that RHB can collide with any of the boxes and make the code compile. That code is reproduced here:

```
#[async_trait(?Send)]
impl Game for WalkTheDog {
    ...
    fn update(&mut self, keystate: &KeyState) {
        ...
        for bounding_box in &walk.platform.bounding_boxes() {
            if walk.boy.bounding_box()
                .intersects(bounding_box) {
                if walk.boy.velocity_y() > 0 &&
                    walk.boy.pos_y() <
                        walk.platform.position.y {
                    walk.boy.land_on(bounding_box.y);
                } else {
                    walk.boy.knock_out();
                }
            }
        }
    }
```

```
                . . .
            }
        }
```

The change here is to loop through all the bounding boxes and check for a collision on any box. There are only three boxes here, so we're not worried about checking all three every time. If the computer can't count to three, you probably need a new computer.

If you temporarily comment out collisions with the stone again, you'll see that you can just barely walk underneath the platform:

Figure 5.28 – Just short enough

At this point, you might wonder whether this should be a collision. After all, his hat does kind of scrape the bottom of the platform. It might be hard for the player to tell if they will fit under the the platform. Can we find a workaround for this?

Game design through constants

As we've gone through this section, we've been introducing more and more constants for values such as FLOOR and PLAYER_HEIGHT. Most of the time, we treat magic numbers as "bad" in code because they lead to duplication and bugs. That's true, but for most of the numbers we've been using, we haven't had duplication. No, in this situation, we can use constants to both clarify what the numbers mean and use those for game design. We can then use game design to hide little quirks, such as our platform being at a height that barely clears the player.

We used `Point { x: 200, y: 400 }` as the location of `Platform` when we originally created it. Those were magic numbers – sorry about that. We actually know that the *y* value of `400` positions the platform at a pretty confusing location. If *y* was `370`, then you would need to go under it, and if it's `420`, you need to go over it. We can create two constants for that and set up the position. That change is shown here:

```
const LOW_PLATFORM: i16 = 420;
const HIGH_PLATFORM: i16 = 375;
#[async_trait(?Send)]
impl Game for WalkTheDog {
    async fn initialize(&self) -> Result<Box<dyn Game>> {
        match self {
            WalkTheDog::Loading => {
                ...
                let platform = Platform::new(
                    platform_sheet.into_serde::<Sheet>()?,
                    engine::load_image("tiles.png").await?,
                    Point {
                        x: 370,
                        y: LOW_PLATFORM,
                    },
                );
```

Figure 5.29 – You're not gonna make it!

You might notice that the platform is a little to the right in this screenshot. I wanted to be able to jump over the rock and then jump onto the platform. It's impossible to do that with the way we constructed it originally, so I moved the platform to the right. I created a constant named `FIRST_PLATFORM` for the *x* location of the platform, set it to `370`, and then set the *x* position of `Platform` to that.

I also found it nearly impossible to actually just jump the stone with the user's combination of speed and gravity. Even after narrowing RHB's collision box, he jumps high but not very far. Fortunately, that was very tweakable with constants – by simply upping `RUNNING_SPEED` from `3` to `4`, he moved quickly enough to make jumping the rock easy.

As we're designing our endless runner, we're going to find that we can hide any imperfections in collisions through game design. You'll constantly need to tweak values such as the speed of the player, bounding box heights, and obstacle locations. The more of the game you can encode into constants, the easier that will be.

A quick challenge

When we wrote the code to cause RHB to get knocked out if he jumps from below the platform, we introduced a bug seen here:

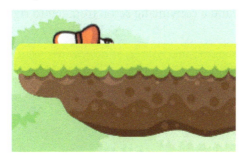

Figure 5.30 – How did he get there?

What's happening is that when RHB collides with the bottom of the platform, he transitions into the `Falling` state, but he doesn't change his velocity, so he continues the jump. Then, gravity stops being applied to RHB in the `KnockedOut` state. Your challenge is to fix this defect. You'll need to modify the states to reflect those changes. Give it a try before you check `https://github.com/PacktPublishing/Game-Development-with-Rust-and-WebAssembly/tree/chapter_5`. The changes are small and all in the existing code.

Summary

In this chapter, we made `WalkTheDog` more closely resemble a game by making RHB run into obstacles and jump onto platforms. We did all of this with axis-aligned bounding boxes and in a scene that looks like a real game, with a background, instead of an empty void. We also dealt with some quirks for dealing with trimmed sprite sheets, properly handled bounding boxes, and utilized the state machine we built in *Chapter 4, Managing Animations with State Machines*, to handle the new animations and manage the state of `RedHatBoy`.

We also learned how collisions are more than just drawing a box around an image. Yes, it's the math behind intersecting boxes, but it's also checking to see whether the player is landing or crashing into the platform. We debugged our collision boxes with rectangles and used those rectangles to make a better fitting box. We even subdivided one image into multiple collision boxes!

This chapter was big, we did a lot, and I encourage you to fiddle with and tweak the code as you see fit. Have RHB jump lower, or higher, or have him move more slowly. Try making the bounding boxes smaller so that it's easier to jump over the stone or put the stone on the platform. Use your imagination!

All in all, we've set up the game so that it's ready to become an endless runner, with randomly generated terrain and a convincing scroll from left to right. We'll develop that in the next chapter.

6
Creating an Endless Runner

Red Hat Boy (**RHB**) can run, jump on a platform, and even crash into a rock and fall over. But once he starts running to his right, he just goes off the screen and is never seen again. There isn't much to it, and if you wait long enough, the game even crashes with a buffer overflow error. In this chapter, we'll make our game truly endless by generating new scenes as RHB runs that contain new obstacles and challenges. They will even contain randomness, and it all starts with RHB staying in one place! It's a real trick.

In this chapter, we will cover the following topics:

- Scrolling the background
- Refactoring for endless running
- Creating a dynamic level

By the end of this chapter, you'll have a functioning endless runner and be able to create obstacles for RHB to hop over and slide under.

Technical requirements

For this chapter, you'll need all the assets at `https://github.com/PacktPublishing/Game-Development-with-Rust-and-WebAssembly/wiki/Assets`. Just like in the previous chapters, you can find the source code at `https://github.com/PacktPublishing/Game-Development-with-Rust-and-WebAssembly/tree/chapter_6`.

Check out the following video to see the Code in Action: `https://bit.ly/35pES1T`

Scrolling the background

To run RHB left to right with an infinite background, we have two choices, as follows:

- Procedurally generate a background, based on a pattern or mathematical formula.
- Use the Hanna-Barbera technique.

While the first option may appear more interesting or dynamic, the Hanna-Barbera technique is much simpler, and it's what we'll using for Walk the Dog. What is the Hanna-Barbera technique? Well, for starters, it may not even go by that name, but that's what I'm calling it. Hanna-Barbera was an animation studio that ran a series of very popular cartoons from the '50s through to the '90s, including Tom and Jerry, The Flintstones, Scooby-Doo, Yogi Bear, and many more. If you were a child in any of those decades, you would wake up to "Saturday morning cartoons," which were dominated by Hanna-Barbera properties. While the studio was known for their beloved characters, they were also known for cutting costs. They produced a lot of cartoons and needed to maximize the ways they could create them quickly and cheaply.

One of the most common traits of Hanna-Barbera cartoons was a repeating background. At the end of an episode of Yogi Bear, for example, Ranger Smith would start chasing Yogi Bear across Jellystone park. However, if you looked closely, Jellystone park appeared to have the same series of trees repeating (see `https://bit.ly/3BNuNXZ` for an example). This money-saving technique is going to work great for our endless runner. We'll use the same background element we're using now and move it to the left as RHB runs to the right. Immediately after, it will become a copy of the same background, making two `Image` elements with the same source image. Once the first image moves completely off screen, we'll move it so that it's to the right of the second image. These two backgrounds will loop, creating the illusion of the background moving forever:

Figure 6.1 – Sliding the canvas over the background

This technique relies on three things. The first is that the background has to be seamless so that there is no visible seam between the two images. Fortunately, our background was built for this and it will work fine. The second is that the canvas window needs to be smaller than the background so that the entire background is never shown on screen. If we do this, then the first background can go entirely off screen to the left and then be moved to the right of the second background, all without any noticeable gaps or tearing. This is because this all happens outside the window's boundaries. I like to think of it as being offstage in a play, then scrambling over to the right-hand side behind the curtain.

Finally, we must use another illusion and freeze the main character in place. Instead of moving the character from left to right on the screen, the objects will move right to left, almost as if on a treadmill. Visually, this will look the same as if the character were running, and it has the advantage of fixing a bug where if the player keeps running right, their x position eventually overflows (becomes bigger than the i16 we are using to hold it) and the game crashes. We'll have to adjust our brains by changing the x velocity from what we expect, but once you get used to it, you'll find that it works quite easily. Let's get started with our scrolling background.

> **Note**
>
> For another example of this technique, go to `https://bit.ly/3BPNBGc`, which explains how this works in a game that continuously moves up, such as Doodle Jump.

Fixing RHB in x

We can scroll the background as much as we want, but if we continue to simultaneously move RHB to the right at the same time, the effect will be having him run at double speed. Instead, we want RHB to run in place while the rocks and platforms move toward him as if they were on a conveyor belt. At the end of this section, we will see RHB run to the right into an empty white void as everything passes past him as if he were running past the end of the world.

Let's start in the `game::red_hat_boy_states` module and not update x in the `update` method of `RedHatBoyContext`:

```
impl RedHatBoyContext {
    fn update(mut self, frame_count: u8) -> Self {
        ...
        // DELETE THIS LINE! self.position.x +=
            self.velocity.x
        self.position.y += self.velocity.y;
        ...
```

With this change, RHB will run in place, with nothing moving around him. We are keeping `velocity` as is because that value is going to be used by the rest of the code base. For ease of use, we'll add a few methods. First, let's add an accessor to the `RedHatBoy` implementation, as shown here:

```
impl RedHatBoy {
    ...
    fn walking_speed(&self) -> i16 {
        self.state_machine.context().velocity.x
    }
```

This function works similar to several of our other accessors for `RedHatBoy`, making it easier to get at the `context` values. Next, let's add a new implementation – `Walk` for the `Walk` struct:

```
impl Walk {
    fn velocity(&self) -> i16 {
        -self.boy.walking_speed()
    }
}
```

The `Walk` implementation is only available when the `WalkTheDog` enum is in the `Loaded` state and it flips `walking_speed` of boy. While boy is moving to the right, this means everything else is moving to the left. Now, in the `update` function of `WalkTheDog`, we can use that value to move everything else to the left. Right after updating `walk.boy`, we can update the `stone` and `platform` positions so that they match the following code:

```
impl Game for WalkTheDog {
    ...
    fn update(&mut self, keystate: &KeyState) {
        if let WalkTheDog::Loaded(walk) = self {
            ...
            walk.boy.update();

            walk.platform.position.x += walk.velocity();
            walk.stone.move_horizontally(walk.velocity());
            ...
```

You should get a compiler error because `stone` doesn't have a `move_horizontally` function. `Stone` is of the `Image` type and can be found in the `engine` module, while `position` on `Image` is private. We'll keep things that way, and instead add `move_horizontally` to the `Image` implementation, as shown here:

```
impl Image {
    ...
    pub fn move_horizontally(&mut self, distance: i16) {
        self.bounding_box.x += distance as f32;
        self.position.x += distance;
    }
}
```

Two things may bother you about this code. The first is that we are directly manipulating `position` on `Platform` but used a method on `Image`. This inconsistency is a *smell* that tells us that something isn't right with our code – in this case, `stone` and `platform` have two different interfaces to modify their positions, even though the code has been duplicated. For now, we'll leave this as is, but it's a hint regarding changes we may want to make later. The other is that we're updating the `bounding_box` and `position` values with the same thing. That's a refactoring we'll leave for the next section (putting a `position` on `Rect Point`), although you can do it now if you're feeling ambitious.

> **Note**
> Code smell is a programming term that was coined by Kent Beck and popularized by Martin Fowler in his book *Refactoring*. If you're getting paid money to program, gaming or not, you should check this book out.

Now, you should see RHB running in place as the rock and platform move beneath him:

Figure 6.2 – Where did the rock go?

> **Tip**
> Don't forget to restart the server if changes don't seem to be showing up. I had to when deleting code, for some reason.

We can start moving the background by matching the `stone` and `platform` movement in the `update` function of `WalkTheDogupdate`. This change will look as follows:

```
fn update(&mut self, keystate: &KeyState) {
    if let WalkTheDog::Loaded(walk) = self {
        ...
        walk.platform.position.x += walk.velocity();
        walk.stone.move_horizontally(walk.velocity());
        walk.background.move_horizontally(walk.velocity());
        ...
```

This small change will mean that RHB can now walk off the edge of the world:

Figure 6.3 – Look, the empty void!

However, we don't want this, so let's learn how to use two tiling backgrounds to simulate an infinite one.

An infinite background

To get an infinite background, we'll need two background images instead of one.
We'll start by storing background as an array instead of just one Image in Walk, as
shown here:

```
struct Walk {
    boy: RedHatBoy,
    backgrounds: [Image; 2],
    stone: Image,
    platform: Platform,
}
```

This will cause several compiler errors because backgrounds doesn't exist; even if it
did, the code expects it to be an Imagearray. Fortunately, the errors largely make sense
and we can figure out what needs to be done. Moving once again to initialize in the
Game implementation, let's set up an array of backgrounds instead of just one when
initializing Walk, as shown here:

```
impl Game for WalkTheDog {

    async fn initialize(&mut self) -> Result<Box<dyn Game>> {
        match self {
            WalkTheDog::Loading => {
                ...
                let background_width = background.width()
                as i16;
                Ok(Box::new(WalkTheDog::Loaded(Walk {
                    boy: rhb,
                    backgrounds: [
                        Image::new(background.clone(),
                        Point { x: 0, y: 0 }),
                        Image::new(
                            background,
                            Point {
                                x: background_width,
                                y: 0,
                            },
                        ),
```

```
        ],
        stone: Image::new(stone, Point { x:
        150, y: 546 }),
        platform,
    }))))
    ...
```

There's a little more going on here compared to our previous changes, so let's go through this code in more detail. The first thing we do is get the width property of background. This is the temporary variable that we created when we loaded HtmlImageElement, not the background property that's attached to Walk that we have been using. We have done this to prevent a borrow-after-move error during the initialization of Walk. Then. we made Walk take an array of Image objects, making sure to clone the background property the first time we create it. Finally, we made sure to position the second Image at background_width so that it will be lined up to the right of the first background, off screen.

However, we still aren't done with compiler errors. This is because the background is being updated and drawn. We'll make the simplest changes we can so that we can start compiling and running again. First, replace the move_horizontally code we just wrote in the update function with the following code, which loops through all the backgrounds and moves them:

```
fn update(&mut self, keystate: &KeyState) {
    if let WalkTheDog::Loaded(walk) = self {
        ...
        walk.platform.position.x += walk.velocity();
        walk.stone.move_horizontally(walk.velocity());
        let velocity = walk.velocity();
        walk.backgrounds.iter_mut().for_each(|background| {
            background.move_horizontally(velocity);
        });
```

Make sure you use iter_mut so that background is mutable. Note that you'll need to bind walk.velocity() to a temporary variable; otherwise, you'll get a compiler error saying cannot borrow '*walk' as immutable because it is also borrowed as mutable. Now, you can update the draw function to draw all the backgrounds:

```
fn draw(&self, renderer: &Renderer) {
```

```
    ...
        if let WalkTheDog::Loaded(walk) = self {
            walk.backgrounds.iter().for_each(|background| {
                background.draw(renderer);
            });

        ...
```

Here, we are looping through backgrounds again and drawing them, relying on the canvas to only show the backgrounds that are on screen. If you play the game while running this code, you'll see that RHB runs farther but doesn't run infinitely. This is because we aren't cycling the backgrounds. If you run the game for long enough, you'll see that the game also crashes with a buffer overflow error, but we'll fix that in the next section. First, we need to get the backgrounds cycling. We can do that by replacing the loop in the update function with code that explicitly destructures the array, as shown here:

```
fn update(&mut self, keystate: &KeyState) {
    if let WalkTheDog::Loaded(walk) = self {
        ...
        walk.platform.position.x += walk.velocity();
        walk.stone.move_horizontally(walk.velocity());

        let velocity = walk.velocity();
        let [first_background, second_background] = &mut
         walk.backgrounds;
        first_background.move_horizontally(velocity);
        second_background.move_horizontally(velocity);

        if first_background.right() < 0 {
            first_background.set_x(
             second_background.right());
        }
        if second_background.right() < 0 {
            second_background.set_x(
             first_background.right());
        }
        ...
```

Here, we start by replacing the `for` loop with `let [first_background, second_background] = &mut walk.backgrounds;` to get access to both backgrounds. Then, we move them both to the left, the same as we did in the loop, and we check whether the right-hand side of the image is negative. This means that the image is off screen, so we can go ahead and move it to the right-hand side of the other background. If you type this in, it won't compile because `set_x` and `right` don't exist on the `Image` struct. Open the `engine` module again so that we can add those to `Image`, as follows:

```
impl Image {
    ...
    pub fn move_horizontally(&mut self, distance: i16) {
        self.set_x(self.position.x + distance);
    }

    pub fn set_x(&mut self, x: i16) {
        self.bounding_box.x = x as f32;
        self.position.x = x;
    }

    pub fn right(&self) -> i16 {
        (self.bounding_box.x + self.bounding_box.width) as
        i16
    }
}
```

Here, we added a `set_x` function that updates `position` and `bounding_box`, just like we did previously, and we had `move_horizontally` call it to avoid duplication. We also added a `right` function that calculates the right-hand side of `bounding_box` based on the current position. With that, RHB now runs to the right, forever! Well, until the buffer overflows and it crashes. Fortunately, we'll take care of that in the next section.

Refactoring for endless running

By now, you've properly noticed a pattern. Every time we add a new feature, we start by refactoring the old code to make it easier to add it. This is generally a good practice in most forms of software development, and we'll be following that same pattern now. We identified a couple of code smells while creating the infinite background, so let's clean those up now, starting with dealing with all those casts.

f32 versus i16

We had to cast values several times to go from `i16` to `f32` and back again. This isn't a safe operation; the maximum of `f32` is orders of magnitude larger than the maximum of `i16`, so there's the potential for our program to crash on a big `f32`. `HtmlImageElement` uses `u32` types, so all the casting to make the compiler shut up isn't even correct. We have two choices here:

- Take our data types (such as `Rect` and `Point`) and make them match `HtmlImageElement`.

- Set `Rect` and any other domain object to be our preferred, smaller, type and cast to the larger type on demand.

I suppose we've been using the second choice so far – that is, cast at random to get the compiler to compile – but that's hardly ideal. While the first option is tempting as we won't have any casts, I prefer `Rect` and `Point` to be as small as possible, so we'll set those up to use `i16` as their values. This is more than large enough for any of our game objects, and the smaller size is potentially beneficial for performance.

> **Note**
>
> The WebAssembly specification does not have an `i32` type, so an `i32` would be just as effective here. It also doesn't have an unsigned type, so it may be worth profiling to see which type is fastest. For our purposes, we'll go with the smallest reasonable size – `i16`. As a professor I once had would say, "We got to the moon on 16 bits!"

To get started with this approach, change all the fields in `engine::Rect` to be `i16` instead of `f32`. Then, follow the compiler errors. Start by getting it to compile, casting `i16` to `f32` as necessary. After getting it to compile and run again, look for anywhere we can cast from `i16` to `f32`, and remove it if possible. This will include looking at the `Land` event in the `Event` enum, which holds an `f32`, and switching it to an `i16`. Finally, look for anywhere you cast to `i16`, and see whetherit's still necessary. It will end up being in a lot of places but it shouldn't be too painful; in the end, there should only be a few necessary casts left. Do this slowly and carefully so that you don't get stuck as you work through the errors.

A more useful Rect

The `Rect` implementation only contains the `intersects` method, but there are two very useful methods it could use: `right` and `bottom`. If you look at the method we just wrote on `Image`, you will see that it's a natural fit for a `right` function. Let's go ahead and add it to `Rect`:

```
impl Rect {
    pub fn intersects(&self, rect: &Rect) -> bool {
        self.x < rect.right()
        && self.right() > rect.x
        && self.y < rect.bottom()
        && self.bottom() > rect.y
    }

    pub fn right(&self) -> i16 {
        self.x + self.width
    }

    pub fn bottom(&self) -> i16 {
        self.y + self.height
    }
}
```

Adding the `right` and `bottom` methods will prevent that addition logic from getting smeared across the game logic. We've also refactored `intersects` to use these new methods. Now, let's go back to the `Image` code we just wrote and update it to use the new `right` method, as shown here:

```
impl Image {
    ...
    pub fn right(&self) -> i16 {
        self.bounding_box.right()
    }
}
```

While we're in `Image`, let's deal with the duplication of `position` and `bounding_box`.

Setting Rect's position

An image containing a bounding_box Rect and a position Point is an accident that occurred due to our code evolving. So, the question is, which one do we want to keep? We could always keep bounding_box for the image, which would mean constructing a Point every time we draw because we need that for the draw_entire_element call. We could also create a Dimens structure that just has width and height, and construct a Rect every time we need it on the update. While I doubt that the cost of creating those objects is going to be noticeable, the fact that it's on every frame is bothersome.

What we'll do instead is give Rect a position field – after all, that's what the x and y coordinates of Rect are. This is a seemingly minor change but with far-reaching implications because we constantly initialize Rect with x and y. Fortunately, we can use the compiler to make this simpler for us. We'll start by changing Rect to hold a position field, instead of x and y:

```
pub struct Rect {
    pub position: Point,
    pub width: i16,
    pub height: i16,
}
```

Adding position is going to cause compiler errors all over the place, as expected. We know that we frequently want to both access the x and y values and create a Rect using x and y, so to make it easier to work with, we'll add two factory methods for Rect, as shown here:

```
impl Rect {
    pub fn new(position: Point, width: i16, height: i16) ->
    Self {
        Rect {
            position,
            width,
            height,
        }
    }

    pub fn new_from_x_y(x: i16, y: i16, width: i16, height:
    i16) -> Self {
        Rect::new(Point { x, y }, width, height)
```

```
    }
    ...
```

Now, when we fix `Rect` everywhere, we will stop creating a `Rect` directly and instead use the new constructor methods. We'll also add getters for x and y because we access those frequently, as shown here:

```
impl Rect {
    ...
    pub fn x(&self) -> i16 {
        self.position.x
    }

    pub fn y(&self) -> i16 {
        self.position.y
    }
```

This gives you most of the tools you will need to fix the compiler errors. I won't reproduce all of them, because there are quite a few and it's repetitive. There are two examples you can use to make take care of all but one error. The first is replacing every reference to `.x` or `.y` with references to the methods.

This is how you do that in the `intersects` method of `Rect`:

```
impl Rect {
    ...
    pub fn intersects(&self, rect: &Rect) -> bool {
        self.x() < self.right()
        && self.right() > rect.x()
        && self.y() < rect.bottom()
        && self.bottom() > rect.y()
    }
```

As you can see, it's the same but with x and y replaced with x() and y(). In addition to seeing errors while accessing x or y, you'll see errors around creating `Rect` because the `position` field isn't specified. You'll want to replace creating `Rect` directly with using one of the constructor methods, as shown here in the implementation of `Image`:

```
impl Image {
    pub fn new(element: HtmlImageElement, position: Point)
```

```
    -> Self {
        let bounding_box = Rect::new(position,
            element.width() as i16, element.height() as i16);
    ...
```

Taking care of those compiler errors, which will show up in both the engine and game modules, will leave you with only one remaining failure. This can be found in the set_x method of Image. This is because we need to set the bounding_box.x value. Rather than using position.x, which will compile but expose us to errors if the internals of Rect change again, we'll add a setter to the Rect implementation, as shown here:

```
impl Rect {
    ...
    pub fn set_x(&mut self, x: i16) {
        self.position.x = x
    }
}
```

Now, in Image, we can fix the last compiler error by using set_x, as shown here:

```
impl Image {
    ...
    pub fn set_x(&mut self, x: i16) {
        self.bounding_box.set_x(x);
        self.position.x = x;
    }
}
```

> **Note**
>
> You may have noticed that the code is inconsistent when it uses setters versus when it uses public variables directly. In general, my rule of thumb is that dumb structures such as Rect don't need setters and getters, especially if we keep them immutable. However, if the internal structure changes, which it did here, then it's time to add an abstraction to hide the internals. This change, from x and y to a position, demonstrated the necessity of the setter after all.

At this point, you should see RHB running to the right and jumping on and off the platform again. Make sure you check out this behavior each time you get to a successful compile since it is easy to make a mistake as you make a large number of small changes.

Now that we've prepared `Rect` to hold a `position`, we can remove the duplication of that data in `Image`. We'll start by removing `position` from the `Image` struct, as shown here:

```
pub struct Image {
    element: HtmlImageElement,
    bounding_box: Rect,
}
```

Now, let's follow the compiler and remove all references to `position` in the `Image` implementation. Fortunately, there are no longer any references to `position` outside of the `Image` implementation, so we can do this by making a few quick changes. These changes are shown here. Note how wherever we previously used `position`, we are now using `bounding_box.position` or `bounding_box.x()`:

```
impl Image {
    pub fn new(element: HtmlImageElement, position: Point)
      -> Self {
        let bounding_box = Rect::new(position,
          element.width() as i16, element.height() as i16);

        Self {
            element,
            bounding_box,
        }
    }

    pub fn draw(&self, renderer: &Renderer) {
        renderer.draw_entire_image(&self.element,
          &self.bounding_box.position)
    }

    pub fn bounding_box(&self) ->&Rect {
        &self.bounding_box
    }

    pub fn move_horizontally(&mut self, distance: i16) {
        self.set_x(self.bounding_box.x() + distance);
```

```
    }

    pub fn set_x(&mut self, x: i16) {
        self.bounding_box.set_x(x);
    }
    ...
```

Now that we've removed the duplication on `Image`, we're ready to get all of the obstacles in a level into one shared `trait` so that we can use them all in one list. Doing that will allow us to fix a bug that occurs when the buffer overflows due to running infinitely and prepare the code for dynamically adding many shared segments. Let's get to it!

Obstacle traits

Currently, the stone and the platform are separate objects on the `Walk` struct. If we want to add more obstacles to the game, we must add more fields to this struct. This is an issue if we want to have an endlessly generated list of things to jump over and slide under. What we'd like to do instead is keep a list of `Obstacles`, go through each one, and check what to do when `RedHatBoy` intersects them. Why do we want to do that? Let's have a look:

- It will eliminate the duplication for knocking out RHB, and eliminate *future* duplication that we'd have to create to continue with our current pattern.

- We want to treat each `Obstacle` as the same so that we can create obstacles on the fly.

- We'll be able to remove any obstacles that have gone off screen.

We'll start by creating an `Obstacle` trait in the `game` module, with one new method named `check_intersection` and two that exist already on `Platform`:

```
pub trait Obstacle {
    fn check_intersection(&self, boy: &mut RedHatBoy);
    fn draw(&self, renderer: &Renderer);
    fn move_horizontally(&mut self, x: i16);
}
```

Why these three methods? `stone` and `platform` are both going to implement `Obstacle`, and we'll need to loop through them, `draw` them, and move them. So, that's why the trait contains `move_horizontally` and `draw`. The new method, `check_intersection`, exists because a `platform` lets you land on it, whereas a `stone` doesn't. So, we'll need an abstraction that can handle intersections differently depending on the type of `Obstacle`. Now that we've created our `trait`, we can implement it on the `Platform` structure. We can start by pulling `draw` out of the `Platform` implementation and creating a `move_horizontally` method, as shown here:

```
impl Obstacle for Platform {
    fn draw(&self, renderer: &Renderer) {
    ...
    }
    fn move_horizontally(&mut self, x: i16) {
        self.position.x += x;
    }
}
```

I've elided the implementation of `draw` here because this method does not change. Meanwhile, `move_horizontally` mimics the code that is currently in `update`, which we identified as a code smell earlier.

Finally, let's add the `check_intersection` function, which currently exists in the `update` method of `WalkTheDog`:

```
for bounding_box in &walk.platform.bounding_boxes() {
    if walk.boy.bounding_box().intersects(bounding_box) {
        if walk.boy.velocity_y() > 0 && walk.boy.pos_y() <
        walk.platform.position.y {
            walk.boy.land_on(bounding_box.y);
        } else {
            walk.boy.knock_out();
        }
    }
}
```

The version that's been implemented for `Platform` should be very similar, without the references to `walk`, as shown here:

```
impl Obstacle for Platform {
    ...
    fn check_intersection(&self, boy: &mut RedHatBoy) {
        if let Some(box_to_land_on) = self
            .bounding_boxes()
            .iter()
            .find(|&bounding_box| boy.bounding_box()
                .intersects(bounding_box))
        {
            if boy.velocity_y() > 0 && boy.pos_y() <
            self.position.y {
                boy.land_on(box_to_land_on.y());
            } else {
                boy.knock_out();
            }
        }
    }
}
```

This code is largely the same but with one fairly significant optimization: instead of looping through every bounding box in `Platform`, this code uses `find` to get the first bounding box that's intersected. If there is one (`if let Some(box_to_land_on)`), then we handle the collision. This prevents redundant checks after a collision is found. The rest of the code is a little bit shorter without the references to `walk`, which is nice. Now, we need to replace `Platform` in `Walk` with a reference to it on the heap, like so:

```
struct Walk {
    boy: RedHatBoy,
    backgrounds: [Image; 2],
    stone: Image,
    platform: Box<dyn Obstacle>,
}
```

> **Note**
>
> We do have an alternative to using a trait object here, which would be using an enum containing every type of obstacle, just like we did with our state machine. The tradeoff to using dynamic dispatch, via the `dyn` keyword, is that a lookup table is stored in memory. The benefit of this is that we write less boilerplate code, and the code doesn't need to be updated every time we add an obstacle. In this case, I think `trait` works better in the same way that an enum works better for a state machine, but it's worth keeping that in mind.

This will cause two compiler errors that we can fix by making small changes. In the `initialize` method of `WalkTheDog`, we are not setting `platform` correctly when we create `Walk`, so let's make a small change, as shown here:

```
impl Game for WalkTheDog {
    async fn initialize(&mut self) -> Result<Box<dyn Game>> {
        match self {
            WalkTheDog::Loading => {
                ...
                Ok(Box::new(WalkTheDog::Loaded(Walk {
                    ...
                    platform: Box::new(platform),
                })))
            }
            ...
```

This is only a one-line change that involves replacing `platform` with `platform: Box::new(platform)`. The other fix is something you'll remember being a smell – setting the position on x directly when `stone` uses a method called `move_horizontally`. This is why we created that method on the `Obstacle` trait on the `Platform` struct. This change can be found in the `update` function for `WalkTheDog`, as shown here:

```
impl Game for WalkTheDog {
    ...
    fn update(&mut self, keystate: &KeyState) {
        if let WalkTheDog::Loaded(walk) = self {
            ...
            let velocity = walk.velocity();
```

```
        walk.platform.move_horizontally(velocity);
        walk.stone.move_horizontally(velocity);
```

Having both platform and stone have a move_horizontally function is a sign that those interfaces can be brought together, which we'll do in a moment. Finally, we must replace the code that we moved into check_intersection with a call to that function. Just a little further down the update function, you'll want to update the following code:

```
impl Game for WalkTheDog {
    ...
    fn update(&mut self, keystate: &KeyState) {
        if let WalkTheDog::Loaded(walk) = self {
            ...
            if second_background.right() < 0 {
                second_background.set_x(
                first_background.right());
            }

            walk.platform.check_intersection(&mut
            walk.boy);

            if walk
                .boy
                .bounding_box()
                .intersects(walk.stone.bounding_box())
            {
                walk.boy.knock_out()
            }
```

The call to check_intersection goes before the check to see whether you've crashed into a stone and after the background updates. You may notice that the code for checking for collisions with a stone is different, in the sense that boy is always knocked out when you collide with it, but is it also conceptually the same because you are, once again, checking for a collision with an obstacle and then doing something. This is why we need to turn stone, which is currently an Image type, into an Obstacle type. But what type should it be?

Barriers versus platforms

We need another type of `Obstacle` that cannot be landed on, and right now a `stone` is an `Image`. Adding features to `Image` isn't appropriate because an `Obstacle trait` is a game concept and `Image` is part of `engine`. Instead, we'll create a type of `Obstacle` that always causes the user to crash into it, called `Barrier`, and turn `stone` into that. It's a very dangerous stone.

We'll start by creating a `Barrier` struct and implementing the `Obstacle` trait with placeholders, as shown here:

```rust
pub struct Barrier {
    image: Image,
}

impl Obstacle for Barrier {
    fn check_intersection(&self, boy: &mut RedHatBoy) {
        todo!()
    }

    fn draw(&self, renderer: &Renderer) {
        todo!()
    }

    fn move_horizontally(&mut self, x: i16) {
        todo!()
    }
}
```

> **Tip**
>
> I generated this skeleton with `rust-analyzer` while using the add-missing-members action. In my editor (emacs), this is as simple as typing c v. In Visual Studio Code, simply click the lightbulb and choose **Implement missing members**. The `todo!` macro throws a runtime exception if this code is called without any implementation, and it is meant to signal temporary code that is there to please the compiler.

> **Note**
>
> Right now, all `Barrier` objects have to be an `Image`, whereas a `Platform` uses a sprite sheet. You may want to use sprite sheets for everything, or even one sprite sheet for everything, and that's fine – better, even. We'll leave things as is here because we've redesigned this application enough already.

Before we fill in all those `todo!` blocks, let's add a typical new method to create the `Barrier` object:

```
impl Barrier {
    pub fn new(image: Image) -> Self {
        Barrier { image }
    }
}
```

Now, we can fill in the functions. The `draw` and `move_horizontally` functions can delegate to `Image`, as shown here:

```
impl Obstacle for Barrier {
    ...

    fn draw(&self, renderer: &Renderer) {
        self.image.draw(renderer);
    }

    fn move_horizontally(&mut self, x: i16) {
        self.image.move_horizontally(x);
    }
}
```

The final function, `check_intersection`, will be a little different. Unlike a `Platform`, which boy can land on, a `Barrier` is always crashed into. The code for this already exists in the `update` method of `WalkTheDog` because it's what we used for `stone`. Let's mimic that implementation here:

```
impl Obstacle for Barrier {
    ...

    fn check_intersection(&self, boy: &mut RedHatBoy) {
        if boy.bounding_box().intersects(
            self.image.bounding_box()) {
```

```
                boy.knock_out()
            }
        }
    }
```

`Barrier` isn't being used anywhere yet. So, we could start by changing `stone` from an `Image` into a `Barrier`. However, we're going to go a little further than that. We're going to create a list in `Walk` that contains *all* the `Obstacle` types. This will let us reduce the amount of specific code in `Walk`, and it will make it far simpler to generate new obstacles on the fly. Remember that's what we're refactoring for. Let's make our list and add it to the `Walk` struct, as shown here:

```
struct Walk {
    boy: RedHatBoy,
    backgrounds: [Image; 2],
    obstacles: Vec<Box<dyn Obstacle>>,
}
```

Note that we've removed `platform` and `stone` from `Walk`, we'll need to update the rest of its implementation and replace direct references to `stone` and `platform` with references to the `Obstacle` vector. This doesn't mean we won't ever mention `platform` and `stone` again; we still have to load the image and sprite sheet, but we'll only mention it once. Once again, we'll look at the compiler error messages, which are complaining a lot about the `initialize`, `update`, and `draw` methods in `WalkTheDog`. Let's start by making changes to the `initialize` function, as shown here:

```
impl Game for WalkTheDog {
    async fn initialize(&self) -> Result<Box<dyn Game>> {
        ...
            Ok(Box::new(WalkTheDog::Loaded(Walk {
                ...
                obstacles: vec![
                    Box::new(Barrier::new(Image::new(
                    stone, Point { x: 150, y: 546 }))),
                    Box::new(platform),
                ],
            })))
        ...
```

We're only changing the construction of the `Walk` construct, replacing the references to `stone` and `platform` by initializing the `obstacles` vector. The first item in the vector is now a `Barrier` but that's just the `stone` object that we created earlier wrapped in the new `Barrier` struct. The second is the `platform` object that we created previously. Everything has to be in a `Box` so that we can use the `Obstacle` trait. The next few changes we'll make must be done in the `update` method. We'll rearrange the code a little bit to update boy first, then our backgrounds, and finally our `obstacles`, as shown here:

```
impl Game for WalkTheDog {

    ...

    fn update(&mut self, keystate: &KeyState) {
        if let WalkTheDog::Loaded(walk) = self {

            ...

            if second_background.right() < 0 {
                second_background.set_x(
                    first_background.right());
            }

            walk.obstacles.iter_mut().for_each(|obstacle| {
                obstacle.move_horizontally(velocity);
                obstacle.check_intersection(&mut walk.boy);
            });
        }
    }

    ...
```

There should be no direct references to `stone` or `platform` in `update`. Now, the code for checking for the movement of obstacles and whether they intersect should only be four lines long and be at the bottom of the `update` method – and that's generously counting the closing brace. Make sure you use the `iter_mut` method since we are mutating `obstacle` in the loop. One of the ways we can tell that we are moving in the right direction in our design is that we're writing *less* code that works with *more* things. Finally, we will need to draw all our `obstacles`, which can be handled by updating the `draw` method, as shown here:

```
impl Game for WalkTheDog {

    ...
```

```
fn draw(&self, renderer: &Renderer) {
    ...

    if let WalkTheDog::Loaded(walk) = self {
        ...
        , walk.obstacles.iter().for_each(|obstacle| {
            obstacle.draw(renderer);
        });
    }
}
}
```

In this case, we can use `for_each` and a plain `iter()`. As you may have guessed, when we want to add more obstacles to the screen, we will just add them to the `obstacles` list. At this point, the code should be working again; RHB should hop his way over a platform and a stone and then crash into it. Now, all we need to take care of is the crash that occurs if we let RHB keep running. We'll handle that next.

Removing obstacles as they go off screen

If you let RHB run to the right for long enough, you'll see a crash message that looks like this:

```
panicked at 'attempt to add with overflow', src/engine.
rs:289:20

Stack:
```

The preceding code is from the log in the browser. Here, the images move farther and farther to the left until they eventually reach the maximum length of the signed 16-bit integer. This is happening because we're never removing an obstacle from the obstacles Vec when they go off screen, and we should. Let's add a line of code to the `update` function that goes right before we move and collide with the obstacles, as shown here:

```
impl Game for WalkTheDog {
    ...

    fn update(&mut self, keystate: &KeyState) {
        if let WalkTheDog::Loaded(walk) = self {
            ...
            walk.obstacles.retain(|obstacle|
```

```
        obstacle.right() > 0);

    walk.obstacles.iter_mut().for_each(|obstacle| {
        obstacle.move_horizontally(velocity);
        obstacle.check_intersection(&mut walk.boy);
    });
    ...
```

The `retain` function will keep any `obstacles` that match the predicate that's been passed in. In this case, this will happen if the rightmost point of the obstacle is to the right of the left edge of the screen. This means we're looping through the list of obstacles twice. If we were using the nightly build of Rust, we could use the `drain_filter` function to avoid that, but our `obstacles` list should never be long enough for that to be an issue. For this code to compile, you'll need to add one more method to the `Obstacle` trait – the `right` method for the rightmost point of `Obstacle`. This can be seen in the following code:

```
trait Obstacle {
    ...
    fn right(&self) -> i16;
}
```

This method will need to be added to both the `Platform` and `Barrier` implementations of `Obstacle`. `Barrier` can just delegate to the image it's holding, `Platform` is a little trickier because it has more than one box. We want to use the right edge of the last bounding box, as shown here:

```
impl Obstacle for Platform {
    ...
    fn right(&self) -> i16 {
        self.bounding_boxes()
            .last()
            .unwrap_or(&Rect::default())
            .right()
    }
}
```

This code gets the last bounding box with `last` and unwraps it since `last` returns an `Option`. We don't want to return a `Result` and then force everybody to use a `Result`, so we are using `unwrap_or(&Rect::default())` to return an empty `Rect` when `Platform` has no bounding boxes. One empty bounding box is effectively the same as no bounding boxes. Then, we get the rightmost value of the last `Rect` with `right`.

`Rect` doesn't have a default implementation yet, so we'll need to add a `#[derive(Default)]` annotation to the `Rect` and `Point` structures in `engine`. The annotation automatically implements the `Default` trait for a `struct` by using the default value of every field in that `struct`. `Point` will need the annotation because it is in the `Rect` structure, so for the macro to work for `Rect`, it must also work for `Point`. Fortunately, there's no real harm in adding this to them.

With that, you can let RHB run for as long as he wants, with no buffer overflow. Now, we need to give RHB many platforms to jump on. We will start by sharing the sprite sheet. Let's dig into this last piece of refactoring.

Sharing a sprite sheet

Each `Platform` has a reference to an `Image` and a `Sheet` that we've casually been referring to as "the sprite sheet." When we start generating more `Platform` objects, we'll want to share a reference to the sheet. So, the time has come to add a `SpriteSheet` struct to our engine to enable that. Let's open the `engine` module and add that new concept.

Creating a sprite sheet

We will start by creating a `struct` that holds both `HtmlImageElement` and `Sheet` in the `engine` module:

```
pub struct SpriteSheet {
    sheet: Sheet,
    image: HtmlImageElement,
}
```

Now, let's create an implementation that will wrap the common behaviors of the sheet that we're using in `Platform`:

```
impl SpriteSheet {
    pub fn new(sheet: Sheet, image: HtmlImageElement) ->
    Self {
        SpriteSheet { sheet, image }
```

```
    }

    pub fn cell(&self, name: &str) -> Option<&Cell> {
        self.sheet.frames.get(name)
    }

    pub fn draw(&self, renderer: &Renderer, source: &Rect,
      destination: &Rect) {
        renderer.draw_image(&self.image, source, destination);
    }
}
```

I initially considered having draw take the name of the cell property we were drawing, but right now, our Platform draws more than one cell at a time, and we want to keep that functionality. Let's replace HtmlImageElement and Sheet in Platform with the SpriteSheet field, as shown here:

```
pub struct Platform {
    sheet: SpriteSheet,
    position: Point,
}
```

Don't forget to import SpriteSheet from the engine module. Now, you can follow the compiler to simplify Platform by removing references to Sheet and HtmlImageElement and just using SpriteSheet. In particular, you'll need to change the new function so that it takes one SpriteSheet instead of the two parameters. The following code shows how this can be initialized in the initialize method of WalkTheDog:

```
impl Game for WalkTheDog {
    async fn initialize(&mut self) -> Result<Box<dyn Game>>
    {
        match self {
            WalkTheDog::Loading => {
                ...
                let platform = Platform::new(
                    SpriteSheet::new(
                        platform_sheet.into_serde::
                        <Sheet>()?,
```

```
        engine::load_image(
            "tiles.png").await?,
    ),
    Point { x: 200, y: 400 },
);
...
```

The rest of Platform can be modified to fit the new interface. Note how you no longer need to say frames and can just call sheet.cell. The draw method will now delegate to self.sheet.draw and pass it the renderer instead of an Image. This structure is small and wouldn't be worth the effort if we didn't want to share the same SpriteSheet across multiple Platform objects. But we do want to share one SpriteSheet, instead of duplicating that memory everywhere. Due to this, we need to make it possible to share it.

Sharing a sprite sheet

To share SpriteSheet across more than one Platform, we'll need to store it somewhere that all of the platforms can point to, and designate something to be the owner of SpriteSheet. We could give SpriteSheet a static lifetime, and make it global, but that would mean making it an Option since it's not available until initialize is used. Instead, we'll store a reference-counted version of SpriteSheet in the Walk structure. This is a tradeoff since we'll be using reference counting instead of ownership to track when we should delete SpriteSheet, but in exchange, we'll only be duplicating the pointer in memory instead of an entire SpriteSheet.

Let's add obstacle_sheet to the Walk struct, as shown here:

```
struct Walk {
    obstacle_sheet: Rc<SpriteSheet>,
    ...
}
```

You'll need to make sure you add use std::rc::Rc to the top of the game module. We'll also need to make sure that Platform can take a reference-counted SpriteSheet instead of taking ownership of SpriteSheet, as shown here:

```
pub struct Platform {
    sheet: Rc<SpriteSheet>,
    ...
}
```

```
impl Platform {
    pub fn new(sheet: Rc<SpriteSheet>, position: Point) ->
    Self {
        Platform { sheet, position }
    }
    ...
```

Here, we're replacing SpriteSheet with Rc<SpriteSheet>. This leaves us with one last modification we need to make – we must initialize the Walk struct and set up obstacle_sheet and the platform, as shown here:

```
#[async_trait(?Send)]
impl Game for WalkTheDog {
    async fn initialize(&mut self) -> Result<Box<dyn Game>>
    {
        match self {
            WalkTheDog::Loading => {
                ...
                let tiles = browser::fetch_json(
                    "tiles.json").await?;
                let sprite_sheet =
                Rc::new(SpriteSheet::new(
                    tiles.into_serde::<Sheet>()?,
                    engine::load_image("tiles.png").await?,
                ));
                let platform = Platform::new(
                    sprite_sheet.clone(),
                    Point {
                        x: FIRST_PLATFORM,
                        y: LOW_PLATFORM,
                    },
                );
                ...
                Ok(Box::new(WalkTheDog::Loaded(Walk {
                    ...
                    obstacles: vec![
```

```
                    Box::new(Barrier::new(Image::new(
                    stone, Point { x: 150, y: 546 }))),
                    Box::new(platform),
                ],
                obstacle_sheet: sprite_sheet,
        })))
```

Two sections change in `initialize`. First, after we call `fetch_json` to get `tiles.`
`json`, we use that to create a reference-counted `SpriteSheet` named `sprite_sheet`
with `Rc::new`. Note that we've replaced `let platform_sheet` with `let tiles`
because that's a better name – it's loading `tiles.json` after all. Then, when we create
`platform` with `Platform::new`, we pass it a clone of the created sprite_sheet.
Previously, this was done inline, but we're going to need `sprite_sheet` again in a minute.

Then, when we're creating the `Walk` struct, we need to pass that created sheet to the
`obstacle_sheet` field. This doesn't need to be cloned because `Walk` is the ultimate
owner of `sprite_sheet`, so `sprite_sheet` can be moved into it. This will increment
the reference counter and will not clone the entire `SpriteSheet`. We will need to
clone `obstacle_sheet` every time we create a `Platform` to ensure the references are
counted correctly, but don't worry about this – the compiler will force us to do this.

With that, we're now ready to reevaluate how our `Platform` object works. Currently, it
can only create one `Platform`, but there's no reason it can't create many things the player
can stand on. We'll want that as we generate levels. We'll do that next.

Many different platforms

The current `Platform` struct assumes it's using the same three cells in the sprite sheet,
including calculating the bounding boxes. So, to allow many kinds of platforms to be used,
we'll need to pass in the cells we want to be rendered from the sheet, and we'll need to pass
in custom bounding boxes for each potential `Platform`. For example, imagine that you
wanted to take the provided tileset (`tiles.json`) and arrange them into a little cliff:

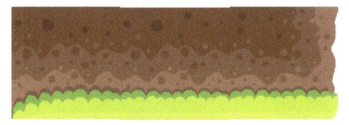

Figure 6.4 – Look out below!

This would require passing the 11, 2, and 3 platform tiles. Those tiles aren't arranged horizontally or neatly, and the bounding boxes don't match our other platform. When we create this platform, we'll need to look up the tile dimensions in tiles.json and work out the bounding boxes from the provided dimensions manually. This means changing the way Platform works so that it's less specific.

Let's start by changing the Platform struct so that it can hold the bounding boxes and a list of the sprites, as shown here:

```
pub struct Platform {
    sheet: Rc<SpriteSheet>,
    bounding_boxes: Vec<Rect>,
    sprites: Vec<Cell>,
    position: Point,
}
```

While we're changing Platform to make it less specific, we're also going to introduce an optimization: Platform will hold the sprite cells instead of looking them up every time they are drawn. There are two optimizations here because we are also storing the bounding boxes for Platform instead of calculating them every time they're created.

This change will break pretty much everything in the implementation of Platform, most notably the new constructor, which will need to take a list of sprite names and bounding boxes and then convert the sprite names into cells, as shown here:

```
impl Platform {
    pub fn new(
        sheet: Rc<SpriteSheet>,
        position: Point,
        sprite_names: &[&str],
        bounding_boxes: &[Rect],
    ) -> Self {
        let sprites = sprite_names
            .iter()
            .filter_map(|sprite_name|
            sheet.cell(sprite_name).cloned())
            .collect();
        ...
```

This isn't the entire new method, just the beginning. We started by changing the signature so that it takes four parameters. `sheet` and `position` were already there but the `new` method now takes a list of sprite names as a reference to an array of string slices. You can take a `Vec` of `String` objects, but it's a lot nicer to use the reference to string slices because it's much easier to call it. Clippy will also object to the code taking a `Vec<String>`, which we will cover in *Chapter 9, Testing, Debugging, and Performance.*

The first thing we do in the constructor is to use an iterator to look up every `Cell` in the sprite sheet via the `filter_map` call. We use `filter_map` instead of `map` because `sheet.cell` can return `None`, so we'll need to skip any invalid sprite names. `filter_map` combines `filter` and `map` to automatically reject any options that have a value of `None` but map the inner value if it is present. The `cloned` method on `Option` will return an `Option<T>` for any `Option<&T>` by cloning the inner value. We use this to take ownership of the inner `Cell`. Let's continue with our constructor:

```
    ...
    let bounding_boxes = bounding_boxes
        .iter()
        .map(|bounding_box| {
            Rect::new_from_x_y(
                bounding_box.x() + position.x,
                bounding_box.y() + position.y,
                bounding_box.width,
                bounding_box.height,
            )
        })
        .collect();

    Platform {
        sheet,
        position,
        sprites,
        bounding_boxes,
    }
}
```

We continue by taking the passed-in bounding boxes, which are of the &[Rect] type, and converting them into a Vec<Rect> to be owned by the Platform struct. However, instead of just calling collect or to_owned, we take each Rect and adjust its position by the actual position of Platform. So, bounding_boxes will need to be passed in relative to its image, where the image starts at (0,0). Imagine that the image you're drawing is positioned in the top-left corner. The bounding boxes are then "drawn" around them, skipping any transparency that's relative to the top-left corner. Then, everything is moved to the right spot in the game. That's the mental model I use to prevent confusion when I'm specifying the bounding boxes later.

> **Note**
> Rust has some pretty good tools for functional-style programming, such as filter and map. It's worth getting to know them.

> **Tip**
> Having four parameters is a lot for a constructor, so you should probably consider replacing this code with the Builder pattern. We did not do this here because it would distract from the topic at hand, but it is a worthwhile code improvement. For an example of this, take a look at the unofficial *Rust Design Patterns* book here: https://bit.ly/3GKxMld.

You'll also need to change the function for retrieving bounding_boxes, which gets a lot smaller:

```
impl Platform {
    ...
    fn bounding_boxes(&self) -> &Vec<Rect> {
        &self.bounding_boxes
    }
}
```

Well, that was a lot easier! Make sure you return a reference to Vec and not a Vec.instance We don't need to make any more calculations here; Platform is being passed its bounding boxes. The rest of the implementation for Platform won't be so easy, as we'll need to modify move_horizontally and draw to account for these changes. The change that needs to be made to move_horizontally is shown here:

```
impl Obstacle for Platform {
    ...
```

```
fn move_horizontally(&mut self, x: i16) {
    self.position.x += x;
    self.bounding_boxes.iter_mut()
    .for_each(|bounding_box| {
        bounding_box.set_x(bounding_box.position.x +
            x);
    });
}
```

The original code only moved position because `bounding_boxes` was calculated on demand. Now that `bounding_boxes` is stored on `Platform`, this needs to be adjusted every time we move `Platform`. Otherwise, you'll have images for `Platform` in one place, the bounding boxes in another, and very strange bugs. Ask me how I know.

Finally, let's update the `draw` function for the new structure. Whereas the original implementation assumed that it was three cells wide and looked up each cell on each draw, the new implementation will loop through every cell and draw it individually. It will also need to account for the width of each cell. So, if the cells are 50 pixels wide, then the first cell will be positioned at 0, the second at 50, and so on:

```
impl Obstacle for Platform {
    ...
    fn draw(&self, renderer: &Renderer) {
        let mut x = 0;
        self.sprites.iter().for_each(|sprite| {
            self.sheet.draw(
                renderer,
                &Rect::new_from_x_y(
                    sprite.frame.x,
                    sprite.frame.y,
                    sprite.frame.w,
                    sprite.frame.h,
                ),
                // Just use position and the standard
                    widths in the tileset
                &Rect::new_from_x_y(
                    self.position.x + x,
                    self.position.y,
```

```
                    sprite.frame.w,
                    sprite.frame.h,
            ),
        );
        x += sprite.frame.w;
    });
}
```

This isn't my favorite code in the world, but it gets the job done. It starts by creating a local, temporary x that will calculate the offset from position for each Cell. Then, it loops through the sprites, drawing each one but adjusting them for both position and x. Note how, in the destination Rect, we advance the x position with self. position.x + x. This ensures each cell is drawn to the right of the previous one. Finally, we calculate the next x position based on the width of cell. This implementation of draw does not use the destination_box method, which means nobody uses it, and you can safely delete it.

> **Note**
>
> This code assumes that width is variable but height is constant and that the sprites move from left to right. Here, a two-level platform would need to be constructed with two platforms.

Platform should now work with any list of sprites that we can construct it with. Now, all we need to do is initialize Platform properly in WalkTheDog::initialize, as shown here:

```
impl Game for WalkTheDog {
    async fn initialize(&mut self) -> Result<Box<dyn Game>>
    {
        match self {
            WalkTheDog::Loading => {
                ...
                let platform = Platform::new(
                    sprite_sheet.clone(),
                    Point {
                        x: FIRST_PLATFORM,
                        y: LOW_PLATFORM,
                    },
```

```
        &["13.png", "14.png", "15.png"],
        &[
            Rect::new_from_x_y(0, 0, 60, 54),
            Rect::new_from_x_y(60, 0, 384 - (60
                * 2), 93),
            Rect::new_from_x_y(384 - 60, 0, 60,
                54),
        ],
    );
    ...
```

With that, `Platform` has been created with two more parameters – the list of tiles and the list of bounding boxes – making up the platform we've had all along. Notice that we can now pass in a simple array of strings for the names of the sprites. This is because we accept the `&[&str]` type as a parameter instead of a `Vec<String>`. You may be wondering where I got the three bounding box rectangles from. After all, previously, we were calculating them in the `bounding_boxes` method, using offsets. I simply looked in `tiles.json` and did the math, factoring in the offsets we used earlier. These are the same measurements as the bounding boxes were when we calculated them. You may also be wondering why these don't use constants, especially after I extolled the virtues of using constants in *Chapter 5, Collision Detection*. That's because we're going to create those in the next section.

At this point, you should be back to where you started – with RHB waiting to jump over a rock. Now, we are ready to create a stream of dynamic segments. At the end of the next section, you'll have the constructs you will need for an endless runner.

Creating a dynamic level

The initial screen we've been looking at for so long, with RHB jumping from a stone onto a platform, is what we're going to call a "segment." It's not a technical term, just a concept we've made up for the sake of generating them. As RHB moves to the right (that is, when all the obstacles move to the left), we'll generate new segments to the right, which is just off screen. We'll create these as segments so that we can control what is generated and how they fit together. Think of it like this: if we generated obstacles at random, then our platforms would look messy and would arrange themselves in an unbeatable fashion, like so:

Figure 6.5 – A truly random level

Instead, what we'll do is create a segment where the first one looks exactly like our one platform and one rock, and have them string together via a "timeline" value that's stored in `Walk`. This timeline will represent the right-hand side of the last segment in x. As that value gets closer to the edge of the screen, we'll generate another new segment and move the timeline back out. With this approach, RHB will be able to run for as long as we like, and we will have the freedom of a level designer. We will be able to create segments that are both easy and hard to navigate, though we'll need to make sure they all interlock and can be beaten. This is the fun part!

Creating one segment

We'll start by taking the introductory screen and creating it as a segment. Let's do this by creating a new file called `segments.rs`, making sure to add `mod segments` to the `lib.rs` file. This module isn't created for the typical software design reasons; usually, it's because `game.rs` is getting pretty long and these segments are closer to being levels than they are true code.

> **Note**
>
> Remember that game.rs can be broken down into a module with separate
> files using a directory with a mod.rs file. We're not doing this here because
> I find it gets harder to explain where new code goes – at least in book form –
> when we have a large number of files. If you are comfortable with doing this,
> then feel free to break this down into smaller chunks.

Each segment will be a function that returns a list of obstacles. Let's create a public
function in segments.rs that returns the same list that the game is initialized with:

```rust
pub fn stone_and_platform(
    stone: HtmlImageElement,
    sprite_sheet: Rc<SpriteSheet>,
    offset_x: i16,
) -> Vec<Box<dyn Obstacle>> {
    const INITIAL_STONE_OFFSET: i16 = 150;
    vec![
        Box::new(Barrier::new(Image::new(
            stone,
            Point {
                x: offset_x + INITIAL_STONE_OFFSET,
                y: STONE_ON_GROUND,
            },
        ))),
        Box::new(create_floating_platform(
            sprite_sheet,
            Point {
                x: offset_x + FIRST_PLATFORM,
                y: LOW_PLATFORM,
            },
        )),
    ]
}
```

Look, constants! We want the segments module to look as data-driven as possible, so we'll be using constants throughout this file. This section of code doesn't compile because the `create_floating_platform` function doesn't exist yet, but it does the same things that the corresponding code in the `initialize` method of `WalkTheDog` does. The only differences are that it uses the `create_floating_platform` function, which doesn't exist, and some constants that also do not exist.

The function itself takes `HtmlImageElement` from `stone` and `Rc<SpriteSheet>` to create `Barrier` and `Platform`, respectively, but it also takes an `offset_x` value. That's because while the first `Barrier` and `Platform` may be at `150` and `200`, respectively, in the future, we'll want those to be that many pixels away from the timeline. It returns a vector of obstacles, which we can use in the `initialize` method of `WalkTheDog` and anywhere else that we generate segments.

> **Information**
>
> You may have noticed that we used an `Rc` for `SpriteSheet` but just take ownership of `HtmlImageElement`, which may need to be cloned when it's called. Nice catch! You may wish to consider making `HtmlImageElement` an `Rc` as well. `HtmlImageElement` is small enough that it's probably fine if we clone it, but it may be worth investigating in *Chapter 9, Testing, Debugging, and Performance.*

Let's continue by creating the function that's missing – that is, `create_floating_platform`:

```
fn create_floating_platform(sprite_sheet: Rc<SpriteSheet>,
position: Point) -> Platform {
    Platform::new(
        sprite_sheet,
        position,
        &FLOATING_PLATFORM_SPRITES,
        &FLOATING_PLATFORM_BOUNDING_BOXES,
    )
}
```

This is a pretty small function in that it just delegates to the `Platform` constructor and passes along important information. As you can see, there are two new constants to go along with the others in `stone_and_platform`. I told you that the constants would come back!

> **Tip**
>
> If you want to use `Rect::new_from_x_y` when you're declaring
> `FLOATING_PLATFORM_BOUNDING_BOXES`, you'll need to declare it and
> `Rect::new` as `pub const fn`.

The rest of the segments module consists of constants and `use` statements. You can infer the values for all the constants from the code we used earlier, or just check out `https://github.com/PacktPublishing/Game-Development-with-Rust-and-WebAssembly/blob/chapter_6/src/segments.rs`. Reproducing that code here would amount to padding. By putting all the values in constants, the code looks increasingly data-driven, with functions just returning the data we want for every segment.

> **Tip**
>
> It's possible to serialize these segments into JSON using `serde` and then read them in from JSON files instead of having the levels be written in Rust code. This is an experiment that you can undertake; I prefer the Rust code version.

Once you've filled in the constants and the `use` statements, you can use the new `stone_and_platform` function in the `initialize` method of `WalkTheDog`. Yeah, that one again. Let's replace the hardcoded list of obstacles with a call to this new function:

```
#[async_trait(?Send)]
impl Game for WalkTheDog {
    async fn initialize(&mut self) -> Result<Box<dyn Game>>
    {
        match self {
            WalkTheDog::Loading => {
                ...
                Ok(Box::new(WalkTheDog::Loaded(Walk {
                    ...
                    obstacles: stone_and_platform(stone,
                    sprite_sheet.clone(), 0),
                    obstacle_sheet: sprite_sheet,
                })))
```

Make sure you import `stone_and_platform` from `segments`! Now that we've got a function to create the initial scene, we can add a timeline and start generating scenes again and again. Let's get started.

> **Tip**
>
> You may have noticed that this puts a circular dependency between `segments` and `game`. You're right. To fix this, take anything that `segments` depends on that is in `game` and put it in another module that both `game` and `segments` depend on. This has been left as an exercise for you.

Adding a timeline

We need to initialize the timeline at the width of a segment. We can calculate this by finding the right-most point in the list of obstacles, and we'll use those cool functional constructs we used earlier. This will be a standalone function that we can keep in the `game` module, which looks like this:

```
fn rightmost(obstacle_list: &Vec<Box<dyn Obstacle>>) -> i16 {
    obstacle_list
        .iter()
        .map(|obstacle| obstacle.right())
        .max_by(|x, y| x.cmp(&y))
        .unwrap_or(0)
}
```

This function goes through a `vec` of `Obstacle` and gets its `right` value. Then, it uses the `max_by` function to figure out the maximum value on the right. Finally, it uses `unwrap_or` because while `max_by` can technically return `None`, if it does that here, then we have completely screwed up and may as well shove all the graphics onto the leftmost part of the screen. Now that we have this function, we can add a `timeline` value to the `Walk` struct, as shown here:

```
struct Walk {
    ...
    stone: HtmlImageElement,
    timeline: i16,
}
```

We also added a reference to `HtmlImageElement` because we'll need that later. We will now initialize `Walk` – yes, we're back in that function again – with `stone` and `timeline`. We'll have to tweak the code slightly to deal with the borrow checker:

```
impl Game for WalkTheDog {
    async fn initialize(&mut self) -> Result<Box<dyn Game>>
```

```
    {
        match self {
            WalkTheDog::Loading => {
                ...
                let starting_obstacles = stone_and_platform
                    (stone.clone(), sprite_sheet.clone(), 0);
                let timeline = rightmost(
                    &starting_obstacles);
                Ok(Box::new(WalkTheDog::Loaded(Walk {
                    ...
                    obstacles: starting_obstacles,
                    obstacle_sheet: sprite_sheet,
                    stone,
                    timeline,
                })))
            }
```

Here, we bind `starting_obstacles` and `timeline` before we initialize `Walk` since
we wouldn't be able to get `timeline` as we've moved `obstacles` already. Note how
we now clone `stone` when we pass it into `stone_and_platform`. We'll need to do
this from now on because each `Barrier` obstacle owns an `Image` and, ultimately, its
`HtmlImageElement`. Finally, we pass `stone` and `timeline` into the `Walk` struct.
Now that we have a `timeline field` we can update it, by moving the rightmost edge
of the generated obstacles to the left on each update, and respond to it by generating more
obstacles as necessary. Our `Canvas` is still `600` pixels wide, so let's say that if there are no
obstacles at the rightmost point past `1000`, we need to generate more.

These changes belong in the `update` method of `WalkTheDog`, at the end of the update
logic:

```
impl Game for WalkTheDog {
    ...
    fn update(&mut self, keystate: &KeyState) {
        if let WalkTheDog::Loaded(walk) = self {
            ...
            walk.obstacles.iter_mut().for_each(|obstacle| {
                obstacle.move_horizontally(velocity);
                obstacle.check_intersection(&mut walk.boy);
            });
```

```
        if walk.timeline < TIMELINE_MINIMUM {
            let mut next_obstacles =
              stone_and_platform(
                walk.stone.clone(),
                walk.obstacle_sheet.clone(),
                walk.timeline + OBSTACLE_BUFFER,
            );

            walk.timeline = rightmost(&next_obstacles);
            walk.obstacles.append(&mut next_obstacles);
        } else {
            walk.timeline += velocity;
        }
    }
```

After moving the obstacles, we check whether walk.timeline is < TIMELINE_
MINIMUM, which is set to 1000 at the top of the module. If it is, we create another
stone_and_platform segment at walk.timeline + OBSTACLE_BUFFER, which
is another constant that's set to 20. Why 20? We needed a little buffer to make sure the
segments weren't right on top of each other, and 20 seemed fine. You could use a larger
number or none at all. Then, we update walk.timeline to the rightmost point of
the new obstacles, and we append those obstacles to the list, ready to be drawn.

If walk.timeline is beyond TIMELINE_MINIMUM, we simply decrease it by RHB's
walking speed until the next update. Upon adding this code, you should see something
similar to the following:

Figure 6.6 – As one platform ends, another beckons

That's right – you have an endless runner! So, how come we're only halfway through this book? Well, our runner is a little dull, seeing as it only has the same two objects over and over again. How about we add some randomness and creativity with multiple segments?

Creating segments

Creating random segments means using the random library to choose a different segment each time one is needed. Let's start by extracting the code we wrote previously into a function, as shown here:

```
impl Game for WalkTheDog {
    ...
    fn update(&mut self, keystate: &KeyState) {
        if let WalkTheDog::Loaded(walk) = self {
            ...
            if walk.timeline < TIMELINE_MINIMUM {
                walk.generate_next_segment()
            } else {
                walk.timeline += velocity;
            }
        }
        ...
    }
}

impl Walk {
    ...
    fn generate_next_segment(&mut self) {
        let mut next_obstacles = stone_and_platform(
            self.stone.clone(),
            self.obstacle_sheet.clone(),
            self.timeline + OBSTACLE_BUFFER,
        );

        self.timeline = rightmost(&next_obstacles);
        self.obstacles.append(&mut next_obstacles);
    }
}
```

> **Information**
>
> WalkTheDog has a bad case of **feature envy**, another one of those code smells we talked about previously. You can read more about this at https://bit.ly/3ytptHA, but as this game gets extended, we'll want to move more code out of WalkTheDog and into Walk.

Now that Walk can generate the next segment, we'll use the random crate from *Chapter 1, Hello WebAssembly*, to choose the next segment. Of course, we only have one segment, so that won't mean much. It looks like this:

```
impl Walk {
    ...
    fn generate_next_segment(&mut self) {
        let mut rng = thread_rng();
        let next_segment = rng.gen_range(0..1);

        let mut next_obstacles = match next_segment {
            0 => stone_and_platform(
                self.stone.clone(),
                self.obstacle_sheet.clone(),
                self.timeline + OBSTACLE_BUFFER,
            ),
            _ =>vec![],
        };
        self.timeline = rightmost(&next_obstacles);
        self.obstacles.append(&mut next_obstacles);
    }
}
```

Don't forget to add use rand::prelude::*; at the top of the file. This generates a random number between 0 and, well, 0. Then, it matches that value and generates the selected segment, which in this case will always be stone_and_platform. There's a default case here, but that's just to quiet the compiler – it can't happen. I'll create a second segment called platform_and_stone that is the same as the first one except it flips the position of stone and platform, and then puts the platform higher by using the HIGH_PLATFORM constant we created earlier. Now, the generate_next_segment function looks like this:

```
impl Walk {
    ...
    fn generate_next_segment(&mut self) {
        let mut rng = thread_rng();
        let next_segment = rng.gen_range(0..2);

        let mut next_obstacles = match next_segment {
            0 => stone_and_platform(
                self.stone.clone(),
                self.obstacle_sheet.clone(),
                self.timeline + OBSTACLE_BUFFER,
            ),
            1 => platform_and_stone(
                self.stone.clone(),
                self.obstacle_sheet.clone(),
                self.timeline + OBSTACLE_BUFFER,
            ),
            _ =>vec![],
        };
        self.timeline = rightmost(&next_obstacles);
        self.obstacles.append(&mut next_obstacles);
    }
}
```

Here, you can see that I get two segments, both of which are called in the same way. Make sure `gen_range` now generates a number from `0` to `2`. Upon running this code, I get to see a new segment:

Figure 6.7 – Who moved that rock?

If you try to copy/paste the preceding code, it won't work since you don't have `platform_and_stone`. This hasn't been included here because you have all the knowledge you need to create your *own* segments. You can start by copying/pasting `stone_and_platform` and tweaking its values. Then, you can try creating platforms with the sprite sheet. Remember that you're not limited to just the three images in our sprite sheet. The entire sheet looks like this:

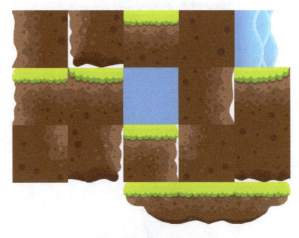

Figure 6.8 – The sprite sheet

You can use this to make larger platforms, steps, and even cliffs. Try making a few different shapes. Try making smaller platforms by skipping the middle tile in the platform we've been using. RHB can slide; can you make something for him to slide under?

For a real challenge, take a look at the water sprites. Currently, RHB can't fall through the ground since we're using a FLOOR variable, but what if we didn't? Could RHB drown? Fall off a cliff, perhaps? It's time to become a game designer!

Summary

It's time for a confession. If you're like me, a programmer, that means you're probably sitting in a room with a bunch of books like this one behind you. Of those books, you've probably only opened half of them, and you've probably only read one or two of them cover-to-cover. Harry Potter notwithstanding.

Great news! At this point, you've made an endless runner. It's got no sound, the collision boxes are pretty big (have you tried to go under a platform yet?), and there's no menu system, but at this point, you have a game. You have the skeleton to make it more fun as you play around, and you're welcome to use this to make even larger or completely different endless runners. I wouldn't hold it against you if you stopped following along at this point, because you've learned a ton.

But if you do decide to stick around for the next chapter, we'll be adding a requirement for immersion that's required for any game – *sound*. Don't you want to hear what RHB sounds like?

7
Sound Effects and Music

Take a moment and think of the game Tetris. If you're like me, you're probably already humming its theme song, *Korobeiniki*, because that song is so synonymous with the game itself. Beyond the appeal of music, sound effects are crucial for creating an immersive experience. We play games with more than just the touch of a keyboard or joystick and the use of our eyes; we hear Mario jump or Sonic catch a ring. While our game may be playable, it's just not a game without some sound. To play sound in our game, we'll need to learn how to use the browser's Web Audio API for both short and long sounds.

In this chapter, we will cover the following topics:

- Adding the Web Audio API to the engine
- Playing sound effects
- Playing long music

By the end of this chapter, you won't just see RHB run, jump, and dodge obstacles, but you'll be able to hear him too after we add sound effects and music to our game. Let's get started!

Technical requirements

The technical requirements are largely unchanged from the previous chapters. You will need the `sound` assets from the `sound` directory in the assets download at `https://github.com/PacktPublishing/Game-Development-with-Rust-and-WebAssembly/wiki/Assets`.

All sounds are from open sound collections and are used with permission. See the `sounds/credits.txt` file for more information. The code for this chapter is available at `https://github.com/PacktPublishing/Game-Development-with-Rust-and-WebAssembly/tree/chapter_7`.

Check out the following video to see the Code in Action: `https://bit.ly/3JUdA2R`

Adding the Web Audio API to the engine

In this section, we'll be using the browser's Web Audio API to add sound to our game. The API is incredibly full-featured, allowing for mixing audio sources and special effects, but we're just going to use it to play background music and sounds. In fact, the Web Audio API is its own book and, if you're interested, you can find one at `https://webaudioapi.com/book/`. While it would be fun to add things such as spatialized audio to our game, we're going to focus on just adding some music and sound effects. I encourage you to experiment on your own when making your own, more complicated games.

Once we've got an overview of the Web Audio API, we'll create a module to play sounds in Rust, load the sounds in the same way as we load our images, and finally, add that sound to the engine.

The Web Audio API is a relatively new technology that is meant to replace older technology for audio, such as QuickTime and Flash, as well as being a more flexible solution than using audio elements. It's supported by all the major browsers, with only old versions of Internet Explorer being a potential problem. Given that the last release of Internet Explorer was in 2013, with Windows using the Edge browser instead, your game is probably okay with sacrificing that market.

The Web Audio API may initially look familiar when compared to Canvas. As with Canvas, you create a context that then provides an API for playing sounds. At that point, the similarity ends. Because the Web Audio API has all the features I mentioned earlier, it can be hard to figure out how to do the basic act of playing a sound. Unlike Canvas, there's no `drawImage` equivalent called `playSound` or something like that. Instead, you have to get the sound data, create `AudioBufferSourceNode`, connect it to a destination, and then finally start it. This enables some really impressive effects (such as the ones found at `https://webaudiodemos.appspot.com/`) but means that, for our game, we'll write the one-time code and forget all about it. In JavaScript, the code to load and prepare a sound for playback looks like the following:

```
const audioContext = new AudioContext();
let sound = await fetch("SFX_Jump_23.mp3");
let soundBuffer = await sound.arrayBuffer();
let decodedArray = await audioContext.
decodeAudioData(soundBuffer);
```

It starts by creating a new `AudioContext`, which is built into the browser engine, then fetching a sound file from the server. The `fetch` call eventually returns a response, which we'll need to decode. We do this by first getting its `arrayBuffer`, which consumes it, and then we use the `audioContext` we created at the beginning to decode the buffer into a sound that can be played. Note how everything is asynchronous, which will cause us a little trouble in the Rust code as we map JavaScript promises to Rust futures. The previous code should only be done *once* for any sound resource since loading and decoding the file can take significant time.

The following code will play a sound:

```
let trackSource = audioContext.createBufferSource();
trackSource.buffer = decodedArray;
trackSource.connect(audioContext.destination);
trackSource.start();
```

Ugh, that's not intuitive, but it's what we have. Fortunately, we can wrap it in a few simple functions that we'll be able to remember, and forget all about it. It creates the `AudioBufferSourceNode` we need with `createBufferSource`, assigns it the array that we decoded into audio data in the previous section, connects it to the `audioContext`, and finally, plays the sound with `start`. It's important to know that you cannot call `start` on `trackSource` twice, but fortunately, the creation of a buffer source is very fast and won't require us to cache it.

That's great! We know the eight lines of code to play a sound in JavaScript, but how do we get this into our engine?

Playing a sound in Rust

We're going to create a `sound` module that's very similar to our `browser` module, a series of functions that just delegate right to the underlying JavaScript. It will be a very bottom-up approach, where we'll create our utility functions and then create the final functions that use them. We'll start by focusing on the parts we need for a `play_sound` function.

> **Note**
>
> Remember that you want these functions to be very small – it's a *thin* layer between Rust and JavaScript – but also to change the interface to better match what you want to do. So, eventually, rather than talking about buffer sources and contexts, we'll want to call that `play_sound` function we wish existed in the first place.

We'll start by creating the module in a file named `sound.rs` living alongside the rest of our modules in `src`. Don't forget to add a reference to it in `src/lib.rs`, as shown here:

```
#[macro_use]
mod browser;
mod engine;
mod game;
mod segments;
mod sound;
```

That's the part I always forget. Our first function will create an `AudioContext` in a *Rusty* way as opposed to the JavaScript way we already saw, and that's as follows:

```
use anyhow::{anyhow, Result};
use web_sys::AudioContext;

pub fn create_audio_context() -> Result<AudioContext> {
    AudioContext::new().map_err(|err| anyhow!
        ("Could not create audio context: {:#?}", err))
}
```

As usual, the Rust version of the code is more verbose than the JavaScript version. That's the price we pay for the positives of Rust. None of this code is particularly new; we're mapping new `AudioContext` to `AudioContext::new`, and we're mapping the `JsResult` error to an `anyhow` result that it might return, to be more Rust-friendly. This code doesn't compile though; take a moment and think about why. It's the infamous feature flags for `web-sys` in `Cargo.toml` that we haven't added `AudioContext` to, so let's add that now:

```
[dependencies.web-sys]
version = "0.3.55"
features = ["console",
            "Window",
            "Document",
            "HtmlCanvasElement",
            "CanvasRenderingContext2d",
            "Element",
            "HtmlImageElement",
            "Response",
            "Performance",
            "KeyboardEvent",
            "AudioContext"
            ]
```

> **Note**
>
> Documentation for the `AudioContext` bindings can be found at https://bit.ly/3tv5PsD. Remember you can search the `web-sys` documentation for any JavaScript object to find its corresponding Rust library.

> **Tip**
>
> Depending on your editor of choice, you may need to restart `rust-analyzer` to get correct compiler errors and code actions when adding a new file to the project (such as `sound.rs`) and/or adding feature flags to the `Cargo.toml` file.

Now that we've set up the sound module, created the function to create AudioContext, and refreshed our memory on the process of adding a new feature to the web-sys dependency, we can go ahead and add a little more code to play sounds. Let's introduce all the remaining feature flags you'll need to add to web-sys in Cargo.toml:

```
[dependencies.web-sys]
version = "0.3.55"
features = ["console",
            "Window",
            "Document",
            "HtmlCanvasElement",
            "CanvasRenderingContext2d",
            "Element",
            "HtmlImageElement",
            "Response",
            "Performance",
            "KeyboardEvent",
            "AudioContext",
            "AudioBuffer",
            "AudioBufferSourceNode",
            "AudioDestinationNode",
            ]
```

The three features, AudioBuffer, AudioBufferSourceNode, and AudioDestinationNode, correspond to those same objects in the original JavaScript code. For instance, the let trackSource = audioContext.createBufferSource(); function returned AudioBufferSourceNode. The web-sys authors have chosen to hide a large number of audio features under individual flags, so we need to name them one at a time.

> **Tip**
> Remember to check the feature flags whenever you can't use a web-sys feature. It's always listed in the documentation with a note such as "This API requires the following crate features to be activated: AudioContext."

Now that we have the features ready, we can add the rest of the code without worrying about those errors. Back in the sound module, the code will look like this:

```
use anyhow::{anyhow, Result};
use web_sys::{AudioBuffer, AudioBufferSourceNode, AudioContext,
AudioDestinationNode, AudioNode};
...
fn create_buffer_source(ctx: &AudioContext) ->
Result<AudioBufferSourceNode> {
    ctx.create_buffer_source()
        .map_err(|err| anyhow!("Error creating buffer
            source {:#?}", err))
}


fn connect_with_audio_node(
    buffer_source: &AudioBufferSourceNode,
    destination: &AudioDestinationNode,
) -> Result<AudioNode> {
    buffer_source
        .connect_with_audio_node(&destination)
        .map_err(|err| anyhow!("Error connecting audio
            source to destination {:#?}", err))
}
```

In this book, we've typically gone through the code one function at a time, but for these two it's not necessary. These functions correspond to the calls to audioContext. createBufferSource and trackSource.connect(audioContext. destionation) respectively. We've converted the code from the object-oriented style of JavaScript into a slightly more procedural format with the functions taking parameters, in part so that we can map errors from the JsValue types into proper Rust Error types via the anyhow! macro.

Now that we have the three functions, we need to play a sound. We can go ahead and write the function that plays it right below them, shown here:

```
pub fn play_sound(ctx: &AudioContext, buffer: &AudioBuffer) ->
Result<()> {
    let track_source = create_buffer_source(ctx)?;
    track_source.set_buffer(Some(&buffer));
```

```
    connect_with_audio_node(&track_source,
    &ctx.destination())?;
        track_source
        .start()
        .map_err(|err| anyhow!
            ("Could not start sound!{:#?}", err))
}
```

The `play_sound` function accepts `AudioContext` and `AudioBuffer` as parameters, then returns the result of the `start` call, with `JsValue` mapped to `Error`. We haven't created an `AudioBuffer` yet anywhere, so don't worry that you don't know how to as we'll cross that bridge when we come to it. What we have here is a function that is very similar to the original JavaScript for playing a sound, but with the additional error handling that comes with Rust, including using the `?` operator to make it easier to read, and a little bit of additional work around `None` in the `track_source.set_buffer(Some(&buffer));` line, where we need to wrap a reference to `AudioBuffer` in `Some` because `track_source` has an optional buffer. In JavaScript, this is `null` or `undefined`, but in Rust, we need to use the `Option` type. Otherwise, both the JavaScript and Rust versions do the same thing to play a sound:

1. Create `AudioBufferSource` from `AudioContext`.

2. Set `AudioBuffer` on the source.

3. Connect `AudioBufferSource` to the `AudioContext` destination.

4. Call `start` to play the sound.

This seems like a lot, but in reality, it's very fast, so there's not much use in caching `AudioBufferSource`, especially since you can only call `start` once. Now that we can play a sound, it's time to load a sound resource and decode it, so that we have an `AudioBuffer` to play. Let's do that now.

Loading the sound

To load a sound from the server, we'll need to translate the following code, which you've already seen, into Rust:

```
let sound = await fetch("SFX_Jump_23.mp3");
let soundBuffer = await sound.arrayBuffer();
let decodedArray = await audioContext.
decodeAudioData(soundBuffer);
```

Fetching the resource is something we can already do in our `browser` module, but we don't have a handy way to get its `arrayBuffer`, so we'll need to add that. We'll also need to create a Rust version of `decodeAudioData`. Let's start with the changes we need to add to `browser`, which are modifications to existing methods. We'll want to split the old `fetch_json` function, which looks like this:

```
pub async fn fetch_json(json_path: &str) -> Result<JsValue> {
    let resp_value = fetch_with_str(json_path).await?;
    let resp: Response = resp_value
        .dyn_into()
        .map_err(|element| anyhow!("Error converting {:#?}
            to Response", element))?;

    JsFuture::from(
        resp.json()
            .map_err(|err| anyhow!("Could not get JSON from
                response {:#?}", err))?,
    )
    .await
    .map_err(|err| anyhow!("error fetching json {:#?}", err
        ))
}
```

We need to split it into two functions that first fetch `Result<Response>`, then a second that converts it into JSON:

```
pub async fn fetch_response(resource: &str) -> Result<Response>
{
    fetch_with_str(resource)
        .await?
        .dyn_into()
        .map_err(|err| anyhow!("error converting fetch to
            Response {:#?}", err))
}

pub async fn fetch_json(json_path: &str) -> Result<JsValue> {
    let resp = fetch_response(json_path).await?;
```

```
JsFuture::from(
    resp.json()
        .map_err(|err| anyhow!("Could not get JSON from
            response {:#?}", err))?,
)
.await
.map_err(|err| anyhow!("error fetching JSON {:#?}", err
    ))
}
```

This is a classic case of *the second person pays for abstraction*, where we wrote the code we needed in *Chapter 2, Drawing Sprites*, to load JSON, but now we need a version of `fetch` that can handle multiple kinds of responses, specifically, sound files that will be accessible as an `ArrayBuffer` instead. That code will need `fetch_response` but will convert it into a different object. Let's write that code now, right below `fetch_json`:

```
pub async fn fetch_array_buffer(resource: &str) ->
Result<ArrayBuffer> {
    let array_buffer = fetch_response(resource)
        .await?
        .array_buffer()
        .map_err(|err| anyhow!("Error loading array buffer
            {:#?}", err))?;

    JsFuture::from(array_buffer)
        .await
        .map_err(|err| anyhow!("Error converting array
            buffer into a future {:#?}", err))?
        .dyn_into()
        .map_err(|err| anyhow!("Error converting raw
            JSValue to ArrayBuffer {:#?}", err))
}
```

Just as `fetch_json` does, this starts by calling `fetch_response` with the passed-in resource. Then, it calls the `array_buffer()` function on that response, which will return a promise that resolves to `ArrayBuffer`. Then, we convert from a promise to `JsFuture` as usual, in order to use the `await` syntax. Finally, we call `dyn_into` to convert the `JsValue` that all `Promise` types return into `ArrayBuffer`. I've skipped over it, but at each step, we use `map_err` to convert the `JsValue` errors into `Error` types.

The `ArrayBuffer` type is a JavaScript type that isn't available to our code yet. It's a core JavaScript type, defined in the ECMAScript standard, and in order to use it directly, we need to add the `js-sys` crate. This is somewhat surprising, as we are already pulling in `wasm-bindgen` and `web-sys`, which are both dependent on JavaScript, so why do we need to pull in yet another crate for `ArrayBuffer`? This has to do with how the various crates are arranged. The `web-sys` crate has all the web APIs where `js-sys` is limited to code that is in the ECMAScript standard. Up to now, we haven't had to use anything in core JavaScript except what was exposed by `web-sys`, but this changes with `ArrayBuffer`.

In order for this code to compile, you'll need to add `js-sys = "0.3.55"` to the list of dependencies in `Cargo.toml`. It is already in `dev-dependencies`, so you can just move it from there. You'll also need to add a `use js_sys::ArrayBuffer` declaration to import the `ArrayBuffer` struct.

> **Tip**
> The various libraries are likely to change in small ways after the publication of this book. If you have any difficulties with these dependencies, check the documentation at `https://github.com/rustwasm/wasm-bindgen`.

Now that we can fetch a sound file and get it as an `ArrayBuffer`, we're ready to write our version of `await audioContext.decodeAudioData(soundBuffer)`. By now, you may have noticed that we're following the same pattern for wrapping every JavaScript function like this:

1. Convert any function that returns a promise, such as `decode_audio_data`, into `JsFuture` so you can use it in asynchronous Rust code.

2. Map any errors from `JsValue` into your own error types; in this case, we're using `anyhow::Result` but you may want more specific errors.

3. Use the `?` operator to propagate errors.

4. Check for feature flags, particularly when using web_sys and you just *know* a library exists.

    To this, we'll add one more step.

5. Cast from JsValue types to more specific types using the dyn_into function.

Following that same pattern, the Rust version of decodeAudioData goes in the sound module, like this:

```
pub async fn decode_audio_data(
    ctx: &AudioContext,
    array_buffer: &ArrayBuffer,
) -> Result<AudioBuffer> {
    JsFuture::from(
        ctx.decode_audio_data(&array_buffer)
            .map_err(|err| anyhow!("Could not decode audio from array buffer {:#?}", err))?,
    )
    .await
    .map_err(|err| anyhow!("Could not convert promise to future {:#?}", err))?
    .dyn_into()
    .map_err(|err| anyhow!("Could not cast into AudioBuffer {:#?}", err))
}
```

You'll need to make sure you add use declarations for js_sys::ArrayBuffer and wasm_bindgen_futures::JsFuture, and also wasm_bindgen::JsCast to bring the dyn_into function into scope. Once again instead of directly calling the method on AudioContext, in this case decodeAudioData, we've created a function that wraps the call. It borrows a reference to AudioContext as the first parameter and takes the ArrayBuffer type as the second parameter. This allows us to encapsulate the mapping of errors and casting of results into a function.

This function then delegates to `ctx.decode_audio_data`, passing it `ArrayBuffer`, but if that's all it did we wouldn't really need it. It then takes any error from `ctx.decode_audio_data` and maps it to `Error` with `anyhow!`; in fact, as you can see, it will ultimately do this at every step in the process, pairing that with the `?` operator to propagate the error. It takes a promise from `decode_audio_data` and creates `JsFuture` from it, then immediately calls `await` to wait for completion, corresponding to the `await` call in JavaScript. After handling any errors converting the promise to `JsFuture`, we use the `dyn_into` function to cast it to `AudioBuffer`, ultimately handling any errors with that as well.

That function is the most complicated of the wrapper functions, so let's reiterate the steps we did when translating from one line of JavaScript to nine lines of Rust:

1. Convert any function that returns a promise into `JsFuture` so you can use it in asynchronous Rust code.

    In this case, `decode_audio_data` returned a promise, and we converted it into `JsFuture` with `JsFuture::from`, then immediately called `await` on it.

2. Map any errors from `JsValue` into your own error type; in this case, we're using `anyhow::Result`, but you may want more specific errors.

    We did this three times, as every call seemed to return a `JsValue` version of the result, adding clarifying language to the error messages.

3. Cast from `JsValue` types to more specific types using the `dyn_into` function.

    We did this to convert the ultimate result of `decode_audio_data` from `JsValue` to `AudioBuffer`, and Rust's compiler could infer the appropriate type from the return value of the function.

4. Don't forget to use the `?` operator to propagate errors; note how this function does that twice.

    We used the `?` operator twice to make the function easier to read.

5. Check for feature flags, particularly when using `web_sys` and you just *know* a library exists.

    `AudioBuffer` is feature flagged, but we added that back at the beginning.

This process is a bit more complicated to explain than it is in practice. For the most part, you can follow the compiler and use tools such as `rust-analyzer` to do things such as automatically add `use` declarations.

Now that we've got all the utilities, we need to play a sound. It's time to add that feature to the `engine` module so our game can use it.

Adding audio to the engine

The functions we just created in the `sound` module could be used by the engine directly via delegation functions, but we don't want to make the game worry about `AudioContext`, `AudioBuffer`, and things like that. Just like `Renderer`, we'll create an `Audio` struct that encapsulates the details of that implementation. We'll also create a `Sound` struct to convert `AudioBuffer` into a friendlier type for the rest of the system. Those will be very small, as shown here:

```
#[derive(Clone)]
pub struct Audio {
    context: AudioContext,
}


#[derive(Clone)]
pub struct Sound {
    buffer: AudioBuffer,
}
```

These structs are added to the bottom of the `engine` module, but they can really be put anywhere in the file. Don't forget to import `AudioContext` and `AudioBuffer`! If you're finding yourself getting confused as `engine` and `game` get larger, you're welcome to break that up into multiple files with a `mod.rs` file and a directory, but to follow along, everything needs to end up in the `engine` module. I'm not going to do that because, while it makes the code a bit easier to navigate, it makes it harder to explain and follow along with. Breaking it up into smaller chunks later is an excellent exercise to make sure you understand the code we're writing.

Now that we have a struct representing `Audio` holding `AudioContext`, and a corresponding `Sound` holding `AudioBuffer`, we can add `impl` to `Audio`, which uses the functions we wrote earlier to play a sound. Now, we'll want to add `impl` to the `Audio` struct to play sounds and load them. Let's start with the load implementation, which is probably the hardest, as seen here:

```
impl Audio {
    pub fn new() -> Result<Self> {
        Ok(Audio {
            context: sound::create_audio_context()?,
        })
    }
```

```
pub async fn load_sound(&self, filename: &str) ->
    Result<Sound> {
    let array_buffer =
        browser::fetch_array_buffer(filename).await?;

    let audio_buffer =
        sound::decode_audio_data(&self.context,
            &array_buffer).await?;

    Ok(Sound {
        buffer: audio_buffer,
    })
    }
}
```

This `impl` will start with two methods, the familiar new method that creates an `Audio` struct with `AudioContext`. Pay attention to the fact that new returns a result in this case, because `create_audio_context` can fail. Then, we have the `load_sound` method, which also returns a result, this time of the `Sound` type, which is only three lines. This is a sign we did something right with the way we organized our functions in the `sound` and `browser` modules, as we can simply call our `fetch_array_buffer` and `decode_audio_data` functions to get `AudioBuffer` and then wrap it in a `Sound` struct. We return a result and propagate errors via ?. If loading a sound was simple, then playing it is easy in this method on the `Audio` implementation:

```
impl Audio {
    ...
    pub fn play_sound(&self, sound: &Sound) -> Result<()> {
        sound::play_sound(&self.context, &sound.buffer)
    }
}
```

For `play_sound`, we really just delegate, passing along `AudioContext` that `Audio` holds and `AudioBuffer` from the passed-in sound.

We've written a module to play sounds in the API, added loading sounds to the browser, and finally created an audio portion of our game engine. That's enough to play a sound effect in the engine; now we need to add it to our game, and here it's going to get complicated.

Playing sound effects

Adding sound effects to our game is a challenge for several reasons:

- Effects must only occur once:

  We'll be adding a sound effect for jumping (*boing!*) and want to make sure that it only happens one time. Fortunately, we have something for that already, our state machine! We can use `RedHatBoyContext` to play a sound when something happens, something like this (don't add it yet):

  ```
  impl RedHatBoyContext {
      ...
      fn play_jump_sound(audio: &Audio) {
          audio.play_sound(self.sound)
      }
  }
  ```

  This leads directly into our second challenge.

- Playing audio on transitions:

  We want to play the sound at the moment of transition, but most transitions won't play a sound. Remember our state machine uses `transition` to transition from one event to another, and while we could pass in the audio there it would only be used by a small portion of the code in that method. It's a code smell, so we won't do that. `RedHatBoyContext` will have to own the audio and the sound. This isn't ideal, we'd prefer there to be only one audio in the system, but that's not workable with our state machine. That leads to our third problem.

- `AudioContext` and `AudioBuffer` are not `Copy`:

  In order to use syntax such as `self.state = self.state.jump();` in the `RedHatBoy` implementation and have each state transition consume `RedHatBoyContext`, we needed `RedHatBoyContext` to be `Copy`. Unfortunately, `AudioContext` and `AudioBuffer` are not `Copy`, which means `Audio` and `Sound` cannot be `Copy` and, therefore, if `RedHatBoyContext` is going to hold audio and a sound, it cannot also be a copy. This stinks, but we can fix it by refactoring `RedHatBoyContext` and `RedHatBoy` to use the `clone` function as needed.

Having `RedHatBoyContext` own an audio means that there will be more than one `Audio` object in the system potentially, where the other will play music. This is redundant but mostly harmless, so it's the solution we'll go with. It gets us moving forward with development, and in the end, the solution works well. When in doubt, choose the solution that ships.

> **Note**
>
> You may wonder why we don't store a reference to `Audio` in `RedHatBoyContext`. Ultimately, Game is static in our engine, and therefore, an `Audio` reference must be guaranteed to live as long as Game if it's stored as a reference on `RedHatBoyContext`.
>
> There are other options, including using the service locator pattern (https://bit.ly/3A4th2f) or passing in the audio into the `update` function as a parameter, but they all take longer to get us to our end goal of playing a sound, which is the real goal of this chapter.

Before we can add a sound effect to the game, we're going to refactor the code to hold an `Audio` element. Then we'll play the sound effect.

Refactoring RedHatBoyContext and RedHatBoy

We're going to prepare `RedHatBoyContext` and `RedHatBoy` to hold audio and a song before we actually do it because that will make it easier to add the sound. Let's start by making `RedHatBoyContext` just `clone`, as shown here:

```
#[derive(Clone)]
struct RedHatBoyContext {
    frame: u8,
    position: Point,
    velocity: Point,
}
```

All we've done is removed the `Copy` trait from the `derive` declaration. This will cause compiler errors on `RedHatBoyStateMachine` and `RedHatBoyState<S>`, which both derive `Copy`, so you'll need to remove that declaration on those structures as well. Once you've done that, you'll see a bunch of errors like this:

```
nerror[E0507]: cannot move out of `self.state` which is behind
a mutable reference
   --> src/game.rs:134:22
```

```
    |
134 |             self.state_machine = self.state_machine.run();
    |                       ^^^^^^^^^^ move occurs because
`self.state` has type `RedHatBoyStateMachine`, which does not
implement the `Copy` trait
```

As expected, the calls to self.state.<method>, where the method takes self, all fail to compile, because RedHatBoyStateMachine doesn't implement Copy anymore. The solution, and we'll do this on every line with this compiler error, is to explicitly clone the state when we want to make the change. Here's the run_right function with the error:

```
impl RedHatBoy {

    ...

    fn run_right(&mut self) {
        self.state_machine = self.state_machine.
            transition(Event::Run);
    }
```

And, here it is with the fix:

```
impl RedHatBoy {

    ...

    fn run_right(&mut self) {
        self.state_machine = self.state_machine
            clone().transition(Event::Run);
    }
```

Perhaps the most teeth-grindingly offensive instance of this is in the transition method, where we will get a move because of the match statement, shown here:

```
impl RedHatBoyStateMachine {
    fn transition(self, event: Event) -> Self {
        match (self, event) {

            ...

            _ => self,
        }

    }
```

The trouble with this section is that `self` is moved into the `match` statement and cannot be returned in the default case. Trying to use `match` and `self` to get around the issue causes all of the typestate methods, such as `land_on` and `knock_out`, to fail because they need to consume `self`. The *cleanest* fix is shown here:

```
impl RedHatBoyStateMachine {
    fn transition(self, event: Event) -> Self {
        match (self.clone(), event) {
            ...
            _ => self,
        }
    }
}
```

It's gross, I admit, but we are able to keep progressing.

> **Tip**
>
> I know what you're thinking – performance! We're cloning on each transition! You're absolutely right, but do you know that the performance is adversely impacted? The first rule of performance is *measure first*, and until we measure this, we don't actually know if the final version of this code is a problem. I spent a lot of time trying to avoid this `clone` call because of performance concerns, and it turned out not to make much of a difference at all. Make it work, then make it fast.

Once you fix that error a few times, you're ready to add the audio and the sound to `RedHatBoyContext`, but what sound will we play?

Adding a sound effect

Using the Web Audio API, we can play any sound format that is supported by the `audio` HTML element, which includes all the common formats of WAV, MP3, MP4, and Ogg. In addition, in 2017, the MP3 license expired, so if you're concerned about that, don't be; you can use MP3 files for sounds without worry.

Since the Web Audio API is compatible with so many audio formats, you can use sound from all over the internet, provided it's released under the appropriate license. The sound effect we'll be using for jumping is available at `https://opengameart.org/content/8-bit-jump-1` and is released under the *Creative Commons public domain* license, so we can use it without concern. You don't need to download that bundle and browse through it, although you can, but the jump sound is already bundled with this book's assets at `https://github.com/PacktPublishing/Game-Development-with-Rust-and-WebAssembly/wiki/Assets` in the sounds directory. The specific file we want is `SFX_Jump_23.mp3`. You'll want to copy that file into the `static` directory of your Rust project so that it will be available for your game.

Now that `RedHatBoyContext` is ready to hold the `Audio` struct, and the `SFX_Jump_23.mp3` file is available to be loaded, we can start adding that code. Start with adding `Audio` and `Sound` to `RedHatBoyContext` as shown here:

```
#[derive(Clone)]
pub struct RedHatBoyContext {
    pub frame: u8,
    pub position: Point,
    pub velocity: Point,
    audio: Audio,
    jump_sound: Sound,
}
```

Remember to add `use` declarations for `Audio` and `Sound` to the `red_hat_boy_states` module. The code will stop compiling because `RedHatBoyContext` is being initialized without `audio` or `jump_sound`, so we'll need to add that. `RedHatBoyContext` is initialized in the new method of the `RedHatBoyState<Idle>` implementation so we'll change that method to take `Audio` and `Sound` objects that we'll pass into `RedHatBoyContext` as shown here:

```
impl RedHatBoyState<Idle> {
    fn new(audio: Audio, jump_sound: Sound) -> Self {
        RedHatBoyState {
            game_object: RedHatBoyContext {
                frame: 0,
                position: Point {
                    x: STARTING_POINT,
                    y: FLOOR,
                },
```

```
                velocity: Point { x: 0, y: 0 },
                audio,
                jump_sound,
            },
            _state: Idle {},
        }
    }
}
```

We could create an Audio object here, but then the new method would need to return Result<Self> and I don't think that's appropriate. This will move the compiler error, because where we call RedHatBoyState<Idle>::new is now wrong. That is in RedHatBoy::new, which can now also take Audio and Sound objects and pass them through.

This leads us to our infamous initialize function in our Game implementation, which fails to compile because it calls RedHatBoy::new without Audio or Sound. This is the appropriate place to load a file, both because it is async and because it returns a result. We'll create an Audio object in initialize, load up the sound we want, and pass it to the RedHatBoy::new function, as shown here:

```
#[async_trait(?Send)]
impl Game for WalkTheDog {
    async fn initialize(&mut self) -> Result<Box<dyn Game>> {
        match self {
            WalkTheDog::Loading => {
                ...
                let audio = Audio::new()?;
                let sound = audio.load_sound
                    ("SFX_Jump_23.mp3").await?;

                let rhb = RedHatBoy::new(
                    sheet,
                    engine::load_image("rhb.png").await?,
                    audio,
                    sound,
                );
                ...
            }
```

This will get the app compiling again, but we don't do anything with audio or sound. Remember that all this work was done because we wanted to make sure the sound is only played *once* when we jump, and the way to ensure that is to put the playing of the sound in the transition from Running to Jumping. Transitions are done in the various From implementations via methods on RedHatBoyContext. Let's write a small function called play_jump_sound on RedHatBoyContext, as shown here:

```
impl RedHatBoyContext {
    ...
    fn play_jump_sound(self) -> Self {
        if let Err(err) = self.audio.play_sound
            (&self.jump_sound) {
                log!("Error playing jump sound {:#?}", err);
        }
        self
    }
}
```

This function is written a little differently than the other transition side effect functions in this implementation, because play_sound returns a result, but in order to be consistent with the other transition methods, play_jump_sound really shouldn't. Fortunately, failing to play a sound, while annoying, isn't fatal, so we'll log the error and continue if the sound couldn't be played. The code now compiles, but we need to add the call to play_jump_sound to the transition. Look for jump on RedHatBoyState<Running> and modify that transition to call play_jump_sound, as shown here:

```
    impl RedHatBoyState<Running> {
        ...
        pub fn jump(self) -> RedHatBoyState<Jumping> {
            RedHatBoyState {
                context: self
                    .context
                    .reset_frame()
                    .set_vertical_velocity(JUMP_SPEED)
                    .play_jump_sound(),
                _state: Jumping {},
            }
        }
    }
```

When this compiles, run the game and you'll see, and hear, RHB jump onto a platform.

Figure 7.1 – Can you hear it?

> **Tip**
> If, like most developers I know, you have 20+ browser tabs open right now, you
> may want to close them. It can slow down the browser's sound playback and
> make the sound timing off.

Now that you've played one sound effect, consider adding more, for example, when RHB
crashes into an obstacle, or lands cleanly, or slides. The choices are up to you! After you've
had a little fun with sound effects, let's add some background music.

Playing long music

You might think that playing music will mean detecting whether the sound is
complete and restarting it. This is probably true for the browser's implementation,
but fortunately, you don't have to do it. The Web Audio API already has a flag on the
`AudioBufferSourceNode` loop that will play the sound on a loop until it is explicitly
stopped. This will make playing background audio rather simple. We can add a flag to the
`play_sound` function in the `sound` module for the `loop` parameter, as shown here:

```
fn create_track_source(ctx: &AudioContext, buffer:
&AudioBuffer) -> Result<AudioBufferSourceNode> {
    let track_source = create_buffer_source(ctx)?;
```

```
        track_source.set_buffer(Some(&buffer));
        connect_with_audio_node(&track_source,
            &ctx.destination())?;
        Ok(track_source)
}
pub enum LOOPING {
    NO,
    YES,
}

pub fn play_sound(ctx: &AudioContext, buffer: &AudioBuffer,
looping: LOOPING) -> Result<()> {
    let track_source = create_track_source(ctx, buffer)?;
    if matches!(looping, LOOPING::YES) {
        track_source.set_loop(true);
    }

    track_source
        .start()
        .map_err(|err| anyhow!("Could not start sound!
            {:#?}", err))
}
```

This starts with the create_track_source function, which is actually a refactoring of the play_sound function. It takes the first three lines of it and extracts them into a separate function for readability. After that, we create a LOOPING enum and use it to check whether we should call set_loop on track_source. You might wonder why we don't just pass bool as the third parameter, and the answer is that it is going to be much easier to read the first line of code shown here than the second:

```
play_sound(ctx, buffer, LOOPING::YES)
play_sound(ctx, buffer, true)
```

Six months from now, when I don't know what that Boolean is for, I'll have to look it up, whereas the version with the enum is obvious. By adding this flag, our program stops compiling because `Audio` in the engine is still calling `play_sound` with two parameters. We can quickly fix that, as shown here:

```
impl Audio {
    ...
    pub fn play_sound(&self, sound: &Sound) -> Result<()> {
        sound::play_sound(&self.context, &sound.buffer,
            sound::LOOPING::NO)
    }
}
```

We'll also add a new method to play background music, which is just playing a sound with looping turned on:

```
impl Audio {
    ...
    pub fn play_looping_sound(&self, sound: &Sound) ->
        Result<()> {
        sound::play_sound(&self.context, &sound.buffer,
            sound::LOOPING::YES)
    }
}
```

I like how the engine has progressively less flexibility than the `sound` module. The `sound` and `browser` modules are wrappers around the browser functionality; the engine provides utilities to help you make a game. Now that the engine provides a way to play background music, we can actually add it to the game. In the assets, there's a second file in the `sounds` directory, `background_song.mp3`, which you can copy into the `static` directory of this project. Once you've done that, we can load and play the background music in our `Game::initialize` function:

```
#[async_trait(?Send)]
impl Game for WalkTheDog {
    async fn initialize(&mut self) -> Result<Box<dyn Game>> {
        match self {
            WalkTheDog::Loading => {
                ...
                let audio = Audio::new()?;
```

314 Sound Effects and Music

```
      let sound = audio.load_sound
          ("SFX_Jump_23.mp3").await?;
      let background_music = audio.load_sound
          ("background_song.mp3").await?;
          audio.play_looping_sound
              (&background_music)?;
      let rhb = RedHatBoy::new(
          sheet,
          engine::load_image("rhb.png").await?,
          audio,
          sound,
      );
      ...
```

> **Tip**
>
> Check out https://gamesounds.xyz/ for royalty-free sounds for
> your games.

Here, we load the second song, background_song.mp3, and play it immediately with
play_looping_sound. On most browsers, you won't hear the music until you click the
canvas to give it focus, so check that if you don't hear anything. One thing to note is that,
even though that sound is going to go out of scope, the browser will happily keep playing
it. We've passed along the song to the browser and it's in charge now. Nothing changes
about the creation of RedHatBoy as the audio is moved into it, and it will eventually be
in charge of playing sound effects for the game.

> **Tip**
>
> You may want to mute your browser while developing, as each time the
> browser refreshes, the song will restart.

There you have it! A proper game with music and sound effects! Now to add a UI, so we
can actually click **New Game** on it.

Summary

In this chapter, you added sounds to your game using the Web Audio API and got an overview of the API itself. The Web Audio API is very broad and has a ton of features, and I'd encourage you to explore it. Your first challenge is to use the `gain` property to change the volume of the music, which is rather loud at the moment. The Web Audio API also supports features such as stereo surround sound and programmatically generated music. Have some fun and try it out!

You also added a new module to the game, and further extended the game engine to support it. We even covered refactoring and made some trade-offs to ensure the game would finish without requiring a time-consuming *ideal* design. I encourage you to take some time to add more sound effects to the game; you have the skills now to make RHB *thud* when he lands or crashes into a rock. Speaking of crashing into rocks, you're probably sick of having to hit *refresh* every time you do that, so in the next chapter, we'll add a small UI with a wonderful **New Game** button.

8
Adding a UI

It may appear that we've developed everything we need for a video game, and to some extent, we have, except for that annoyance where we need to hit refresh every time little **Red Hat Boy** (**RHB**) hits a rock. A real game has buttons for a "new game" or "high scores", and in this chapter, we'll be adding that UI. To do so may seem trivial, but event-driven UIs that you might be familiar with from web development are an odd fit with our game loop. To add a simple button, we'll need to make significant changes to our application and even write a little HTML.

In this chapter, you'll do the following:

- Design a new game button
- Show the button on game over
- Start a new game

At the end of the chapter, you'll have the framework in place for a more full-featured UI and the skills to make it work.

Technical requirements

You'll need a few more assets, this time from the `ui` directory in the `assets` download at `https://github.com/PacktPublishing/Game-Development-with-Rust-and-WebAssembly/wiki/Assets`. The font is Kenny Future Narrow from `https://www.kenney.nl`. The button is from `https://www.gameart2d.com/`. Both are CC0-licensed. As before, the final code for this chapter is available on the corresponding branch at `https://github.com/PacktPublishing/Game-Development-with-Rust-and-WebAssembly/tree/chapter_8`.

Check out the following video to see the Code in Action: `https://bit.ly/3DrEeNO`

Design a new game button

When RHB crashes into a rock, he falls over and… well, let's say he takes a nap. Unfortunately, at that point, the player has to refresh the page to start a new game. In most games, we'd see a series of buttons for a new game and high scores. For now, we'll just put in a new game button that will restart from the beginning. This might seem like a simple task, but in fact, we'll have quite a bit to do.

First, we need to decide how we want to implement the button. We really have two choices. We can create a button in the engine, which would be a sprite that is rendered to the canvas, the same as everything else, or we can use an HTML button and position it over the canvas. The first option will look right and won't require any traditional web programming, but it will also require us to detect mouse clicks and handle a button-click animation. In other words, we'd have to implement a button. That's more than we want to implement to get our game working, so we're going to use a traditional HTML button and make it *look* like it's a game element.

So, we're going to write some HTML and CSS, which we can use to make the button look like it's a part of the game engine. Then, we'll add the button to the screen with Rust and handle the click event. That will be the tough part.

Preparing a UI

Conceptually, our UI will work like a HUD in a FPS or where a button is superimposed over the front of a game itself. Imagine that there is a perfectly clear pane of glass on top of the game, and the button is a sticker that's stuck to it. This means, in the context of a web page, that we need a div that is the same size and in the same place as the canvas.

> **Tip**
>
> This isn't a book on HTML or CSS, so I'm not going to spend much time covering it, other than the canvas we've been using throughout. If web development isn't your forte, don't worry – a quick scan of `https://learnxinyminutes.com/docs/html/` will cover more than enough. We'll also be using a little bit of CSS in this section, and you can get a similar cheat sheet for that syntax at `https://learnxinyminutes.com/docs/css/`.

We can start rather quickly by updating `index.html` to have the required div, as follows:

```html
<body>
  <div id="ui" style="position: absolute"></div>
  <canvas id="canvas" tabindex="0" height="600" width="600">
    Your browser does not support the Canvas.
  </canvas>
  <script src="index.js"></script>
</body>
</html>
```

Note that the `ui` div is `position: absolute` so that it doesn't "push" the `canvas` element below it. You can see how this will work by putting a standard HTML button in the `div` element, as follows:

```html
<div id="ui" style="position: absolute">
  <button>New Game</button>
</div>
```

This will produce a screen that looks like the following:

New Game

Figure 8.1 – A New Game button!

It won't respond very well if you completely shrink the screen horizontally, but the game won't work in that situation, so it should be fine. Now that we have a button, we'll need to make it look like a game element, and for that, we'll need styling. Go ahead and create a file named `styles.css` in the `static` directory, and add a link to it in `index.html`, as follows:

```html
<html>
<head>
  <meta charset="UTF-8">
    <title>My Rust + Webpack project!</title>
    <link rel="stylesheet" href="styles.css" type="text/css"
      media="screen">
</head>
```

Of course, a link to an empty file doesn't do much for us. To prove the link is working, go ahead and change the `index.html` file slightly, removing the inline style on the `ui` div so that it looks like `<div id="ui">`. This will cause the button to push the canvas element down, and your game will likely be slightly off:

New Game

Figure 8.2 – New Game on top

Now, in the CSS file, you'll want to add a style for that div. It's not really important that this style isn't an inline one, except that this handily checks that our CSS file is being loaded. In the CSS file, insert the following:

```
#ui {
    position: absolute;
}
```

This is a CSS selector for any elements with the `ui` ID and sets their position to `absolute`. If your CSS file is being loaded, then the new game button should be over the top of the canvas again. Later, we'll programmatically add that button in our game code, but for now, we just want it to show up and look right. We'll want to give it a font that looks like a video game, and a background too. Let's start with the font. In your assets, you'll see there is a directory called `ui`, which contains a file named `kenney_future_narrow-webfont.woff2`. **WOFF** stands for **Web Open Font Format** and is a font format that will work in every modern browser.

> **Note**
> Whenever you're unsure whether a feature will work with a browser, and sometimes when you are sure, check `https://caniuse.com/` to double-check. For WOFF files, you can see the results here: `https://caniuse.com/?search=woff`.

Copy `kenney_future_narrow-webfont.woff2` into the `static` directory in your application so that it gets picked up by the build process. Then, you need to specify `@font-face` in CSS so that elements can be rendered in it, which looks like so:

```
@font-face {
  font-family: 'Ken Future';
  src: url('kenney_future_narrow-webfont.woff2');
}
```

What we've done here is load a new font face with the simple name `'Ken Future'` so that we can reference it in other styles, and loaded it via the specified URL. Now, we can change all buttons to use that font with this additional CSS:

```
button {
    font-family: 'Ken Future';
}
```

Now, you should see the button being rendered with a font that looks more like a game, as shown here:

Figure 8.3 – New Game with the Kenney Future Font

The button still looks a lot like an HTML button because of that traditional web background. To make it look more like a game button, we'll use a background and CSS Sprites to create a pretty button with rounded corners and hover colors.

CSS Sprites

As a game developer, you already know what a sprite is; you haven't forgotten *Chapter 2, Drawing Sprites*, already, have you? In the case of **CSS Sprites**, the term as commonly used is a bit of a misnomer, as instead of referring to a sprite, it really refers to a sprite sheet.

Conceptually, CSS Sprites work the same way as rendering them with the canvas. You slice out a chunk of a larger sprite and only render that portion. We'll just do the entire thing in CSS instead of Rust. Since we're using CSS, we can change the background when the mouse is over the button and when it is clicked. This will make the button look correct, and we won't have to write Rust code to have the same effect. Clicking a button is something a browser is very good at, so we'll leverage it.

We'll use the `Button.svg` file from the `ui` directory in the downloaded assets, so you can copy that file to the `static` directory in your game's project. The SVG file actually contains an entire library of buttons, which looks like this:

Figure 8.4 – The top of Button.svg

We'll want to slice out the wide blue, green, and yellow buttons to be the background for the button in various states. We'll start by using the `background` attribute in CSS to set the button's background to the SVG file. You'll update the style as follows:

```
button {
    font-family: 'Ken Future';
    background: -72px -60px url('Button.svg');
}
```

The pixel values in `background`, `-72px` and `-60px`, mean taking the background and shifting it 72 pixels to the left and 60 pixels upward to line it up with the blank blue button. You can get those values in a vector graphics editor such as **Inkscape**. The `url` value specifies which file to load. Make those changes, and you'll see the button change to have a new background… well, sort of.

Figure 8.5 – The button, but with a cut-off background

As you can see, the background is cut off, so you only get half of it, and the button itself still has some of the effects of a default HTML button. We can get rid of those effects with a little more CSS to remove the border and resize the button to match the background, as shown here:

```
button {
    font-family: 'Ken Future';
    background: -72px -60px url('Button.svg');
    border: none;
    width: 82px;
    height: 33px;
}
```

The `width` and `height` values were plucked from *Inkscape~ again, and that will set the button to be the same size as the button background in the source. As with the sprite sheets we used earlier, we need to cut out a slice from the original source, so in this case, there is a rectangle starting at (72, 60) with a width and height of 82x33. With those changes, the button now looks like a game button instead of a web button.

Figure 8.6 – A New Game button

There are still a few problems. The button now doesn't visually interact with the user, so it just looks like a picture when you click it. We can address that with CSS pseudo-classes for `#active` and `#hover`.

> **Note**
> Some browsers, notably Firefox, will render **New Game** on one line instead of two.

> **Note**
>
> For more information on pseudo-classes, check the Mozilla documentation here: `https://developer.mozilla.org/en-US/docs/Web/CSS/Pseudo-classes`.

In each pseudo-class, we'll change the background attribute to line up with another background. Again, the numbers were pulled out of Inkscape, with a little tweaking once they were added to make sure that they lined up. First, we can handle the `hover` style, which is when the mouse is over the image.

That produces a hover button that looks like this:

```
button:hover {
    background: -158px -60px url('Button.svg');
}
```

Figure 8.7 – Hover

Then, we'll add the `active` style, which is what the mouse will look like when clicked:

```
button:active {
    background: -244px -60px url('Button.svg');
}
```

That produces a clicked button like this:

Figure 8.8 – Active

The final issue is that our button is really small, for a game anyway, and is positioned at the upper-left corner. Making the button larger the traditional CSS way with width and height is problematic, as shown here when we change the width value:

Figure 8.9 – That is not a button

Changing the width or height will mean changing the "slice" that we're taking from the sprite sheet, so we don't want that. What we'll use instead is the CSS `translate` property, with the `scale` function, which looks like so:

```
button {
    font-family: 'Ken Future';
    background: -72px -60px url('Button.svg');
    border: none;
    width: 82px;
    height: 33px;
    transform: scale(1.8);
}
```

This gives us a nice large button with the right background, but it's not in the right spot.

Figure 8.10 – The button with the left side cut off

Now that the button is large and looks like a game button, we just need to put it in the right spot. You can do that by adding `translate` to the `transform` property, where `translate` is a fancy way of saying `move`. You can see that as follows:

```
button {
    font-family: 'Ken Future';
    background: -72px -60px url('Button.svg');
    border: none;
    width: 82px;
    height: 33px;
    transform: scale(1.8) translate(150px, 100px);
}
```

This will get the new game button into, roughly, the center of the screen.

Figure 8.11 – A New Game button!

> **Note**
>
> Centering a button in a div requires a little more CSS than I want to cover
> in this book. Since we're positioning things manually, we can go with "good
> enough" for now. If you're more comfortable with web development, feel free to
> make it truly perfectly centered. If you're interested in getting the perfect center
> with Flexbox, take a look here: `https://webdesign.tutsplus.`
> `com/tutorials/how-to-create-perfectly-centered-`
> `text-with-flexbox--cms-27989`.

The new game button now shows up, but it doesn't do anything because our code
isn't doing anything with `onclick`. It's just a floating button, taunting us with its
ineffectiveness. Go ahead and remove the `button` element from `index.html`, but
keep `div` with the `ui` ID. Instead, we'll use Rust to dynamically add and remove the
button when we need it and actually handle the clicks. For that, we'll want to make some
additions to our `browser` and `engine` modules, so let's dig in.

Showing the button with Rust

We've written HTML to show the button and it looks pretty good, but we'll actually need to show it and hide it on command. This means interacting with the browser and using the `browser` module. We haven't done this in a while, so let's refresh our memory on how we translate from the JavaScript we'd write traditionally to the Rust with `web-sys` that we'll be using. First, we'll need code to insert the button into the `ui` div. There are lots of ways to do this; we'll use `insertAdjacentHTML` so that we can just send a string from our code to the screen. In JavaScript, that looks like this:

```javascript
let ui = document.getElementById("ui");
ui.insertAdjacentHTML("afterbegin", "<button>New Game</button>");
```

> **Note**
>
> You can find the docs for this function at `https://developer.mozilla.org/en-US/docs/Web/API/Element/insertAdjacentHTML`. When it comes to looking up browser APIs, the **Mozilla Developer Network (MDN)** is your friend.

We spent a lot of time translating this kind of code into Rust in *Chapter 2, Drawing Sprites*, and *Chapter 3, Creating a Game Loop*, but let's refresh our memory and appease any monsters who read books out of order. Any JavaScript function or method is likely to be found in the `web-sys` crate with the name converted from PascalCase to snake_case, and with most of the functions returning `Option`. Frequently, you can just try that out, and it will work. Let's create a new function in `browser` and see whether that's the case, as shown here:

```rust
pub fn draw_ui(html: &str) -> Result<()> {
    document()
        .and_then(|doc| {
            doc.get_element_by_id("ui")
                .ok_or_else(|| anyhow!("UI element not found"))
        })
        .and_then(|ui| {
            ui.insert_adjacent_html("afterbegin", html)
                .map_err(|err| anyhow!("Could not insert
                    html {:#?}", err))
        })
}
```

This `draw_ui` function assumes there is a div with the `ui` ID, just as the `canvas` function assumes an ID of `canvas`. This means it's not *incredibly* generic, but we don't need a more complex solution right now. If we do later, we'll write more functions. As always, we don't want to go too far with some idea of "perfect" code because we've got a game to finish.

Once again, the Rust version of the code is much longer, using `and_then` and mapping errors to make sure we handle the error cases instead of just crashing or halting the program as JavaScript would. This is another case where code is aesthetically less pleasing in Rust but, in my opinion, better because it highlights the possible causes of an error. The other function we'll need right away is used to hide the `ui` element, which looks like this in JavaScript:

```
let ui = document.getElementById("ui");
let firstChild = ui.firstChild;
ui.removeChild(firstChild);
```

This function grabs the first child of the `ui` div and removes it with the `removeChild` method. To be completely thorough, we should loop through all the `ui` children and make sure they all get removed, but we don't do that here because we already know there's only one. We also remove the children (and don't just set their visibility to hidden) so that they do not affect the layout, and any event listeners are removed. Once again, you'll want to translate JavaScript to Rust. In this case, `firstChild` becomes the `first_child()` method and `removeChild` becomes `remove_child`, as shown here:

```
pub fn hide_ui() -> Result<()> {
    let ui = document().and_then(|doc| {
        doc.get_element_by_id("ui")
            .ok_or_else(|| anyhow!("UI element not found"))
    })?;

    if let Some(child) = ui.first_child() {
        ui.remove_child(&child)
            .map(|_removed_child| ())
            .map_err(|err| anyhow!("Failed to remove child
                {:#?}", err))
    } else {
        Ok(())
    }
}
```

This function is a little different than `draw_ui`, in part because `first_child` being missing isn't an error; it just means you called `hide_ui` on an empty UI, and we don't want that to error. That's why we use the `if let` construct and just return an `Ok(())` explicitly if it isn't present. The `ui` div was already empty, so it's fine. In addition, there's that weird call to `map(|_removed_child| ())`, which we call because `remove_child` returns the `Element` being removed. We don't care about it here, so we are, once again, explicitly mapping it to our expected value of unit. Finally, of course, we address the error with `anyhow!`.

This function reveals some duplication, so let's go ahead and refactor it out in the final version, as follows:

```
pub fn draw_ui(html: &str) -> Result<()> {
    find_ui()?
        .insert_adjacent_html("afterbegin", html)
        .map_err(|err| anyhow!("Could not insert html
            {:#?}", err))
}
pub fn hide_ui() -> Result<()> {
    let ui = find_ui()?;

    if let Some(child) = ui.first_child() {
        ui.remove_child(&child)
            .map(|_removed_child| ())
            .map_err(|err| anyhow!("Failed to remove child
                {:#?}", err))
    } else {
        Ok(())
    }
}
fn find_ui() -> Result<Element> {
    document().and_then(|doc| {
        doc.get_element_by_id("ui")
            .ok_or_else(|| anyhow!("UI element not found"))
    })
}
```

Here, we've replaced both of the repetitive document().and_then calls with calls to find_ui, which is a private function that ensures we always get the same error when UI isn't found. It streamlines a little bit of code and makes it possible to use the try operator in draw_ui. The find_ui function returns Element, so you need to make sure to import web_sys::Element.

We've got the tools we need to draw the button set up in browser. To show our button programmatically, we can just call browser::draw_ui("<button>New Game</button>"). That's great, but we can't actually handle doing anything on the button click yet. We have two choices. The first is to create the button with an onclick handler such as browser::draw_ui("<button onclick='myfunc'>New Game</button>"). This will require taking a function in our Rust package and exposing it to the browser. It would also require some sort of global variable that the function could operate on. If myfunc is going to operate on the game state, then it needs access to the game state. We could use something such as an event queue here, and that's a viable approach, but it's not what we'll be doing.

What we're going to do instead is set the onclick variable in Rust code, via the web-sys library, to a closure that writes to a channel. Other code can listen to this channel and see whether a click event has happened. This code will be very similar to the code we wrote in *Chapter 3*, *Creating a Game Loop*, for handling keyboard input. We'll start with a function in the engine module that takes HtmlElement and returns UnboundedReceiver, as shown here:

```
pub fn add_click_handler(elem: HtmlElement) ->
UnboundedReceiver<()> {
    let (click_sender, click_receiver) = unbounded();
    click_receiver
}
```

Don't forget to bring HtmlElement into scope with use web_sys::HtmlElement. This doesn't do much, and it sure doesn't seem to have anything to do with a click, and it's not obvious why we need an UnboundedReceiver. When we add a click handler to the button, we don't want to have to move anything about the game *into* the closure. Using a channel here lets us encapsulate the handling of the click and separate it from the reacting to click event. Let's continue by creating the on_click handler, as shown here:

```
pub fn add_click_handler(elem: HtmlElement) ->
    UnboundedReceiver<()> {
    let (mut click_sender, click_receiver) = unbounded();
    let on_click = browser::closure_wrap(Box::new(move || {
```

```
        click_sender.start_send(());
    }) as Box<dyn FnMut()>);
    click_receiver
}
```

The changes we've made are to make `click_sender` mutable and then move it into the newly created closure called `on_click`. You may remember `closure_wrap` from the earlier chapters, which needs to take a heap-allocated closure, in other words a `Box`, which, in this case, will be passed a `mouse` event that we're not using so we can safely skip it. The casting to `Box<dyn FnMut()>` is necessary to appease the compiler and allow this function to be converted into `WasmClosure`. Inside that, we call the sender's `start_send` function and pass it a unit. Since we're not using any other parameters, we can just have the receiver check for any event.

Finally, we'll need to take this closure and assign it to the `on_click` method on `elem` so that the button actually handles it, which looks as follows:

```
pub fn add_click_handler(elem: HtmlElement) ->
UnboundedReceiver<()> {
    let (mut click_sender, click_receiver) = unbounded();
    let on_click = browser::closure_wrap(Box::new(move || {
        click_sender.start_send(());
    }) as Box<dyn FnMut()>);
    elem.set_onclick(Some(on_click.as_ref().unchecked_ref()));
    on_click.forget();
    click_receiver
}
```

We've added the call to `elem.set_onclick`, which corresponds to `elem.onclick =` in JavaScript. Note how we pass `set_onclick` a `Some` variant because `onclick` itself can be `null` or `undefined` in JavaScript and, therefore, can be `None` in Rust and is an `Option` type. We then pass it `on_click.as_ref().unchecked_ref()`, which is the pattern we've used several times to turn `Closure` into a function that `web-sys` can use.

Finally, we also make sure to forget the `on_click` handler. Without this, when we actually make this callback, the program will crash because `on_click` hasn't been properly handed off to JavaScript. We've done this a few times, so I won't belabor the point here. Now that we've written all the code, we'll need to show a button and handle the response to it, and we need to integrate it into our game. Let's figure out how to show the button.

Show the button on game over

We can show and hide the button in the `Game` `update` method by checking on each frame if the game is over and if the button is present, ensuring that we only show or hide it once, and that would probably work, but I think you can sense the spaghetti code beginning to form if we do that. In general, it's best to avoid too much conditional logic in `update`, as it gets confusing and allows for logic bugs. Instead, we can think of every conditional check that looks like `if` `(state_is_true)` as two different states of the system. So, if the new game button is shown, that's one game state, and if it isn't, that's another game state. You know what that means – it's time for a state machine.

A state machine review

In *Chapter 4*, *Managing Animations with State Machines*, we converted RHB to a state machine in order to make it change animations on events easily and, more importantly, correctly. For instance, when we wanted RHB to jump, we went from `Running` to `Jumping` via a typestate method, only changing the state one time and changing the velocity and playing the sound one time. That code is reproduced here for clarity:

```
impl RedHatBoyState<Running> {

    ...

    pub fn jump(self) -> RedHatBoyState<Jumping> {
        RedHatBoyState {
            context: self
                .context
                .reset_frame()
                .set_vertical_velocity(JUMP_SPEED)
                .play_jump_sound(),
            _state: Jumping {},
        }
    }
}
```

The typestates work great, but they are also noisy if we don't need that kind of functionality. That's why in that same chapter, we chose to model our game itself as a simple enum, like so:

```
pub enum WalkTheDog {
    Loading,
    Loaded(Walk),
}
```

This is going to change significantly because we now have a problem that necessitates a state machine. When RHB is knocked out, the game is over, and the new game button should appear. That's a side effect that needs to happen once, on a change of state, the perfect use case for our state machine. Unfortunately, refactoring to a state machine is going to require a not insignificant amount of code because our current method for implementing state machines is elegant but a little noisy. In addition, there's actually two state machines at work here, which is not obvious at first. The first is the one we see at the beginning, moving from `Loading` to `Loaded`, which you can think of as when you don't have `Walk` and when you do. The second is the state machine of `Walk` itself, which moves from `Ready` to `Walking` to `GameOver`. You can visualize it like this:

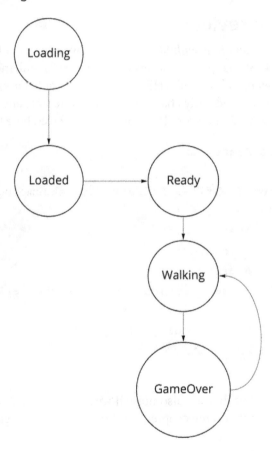

Figure 8.12 – Nested state machines

As you can see, we have two state machines here, one going from `Loading` to `Loaded` and the other representing the three game states of `Ready`, `Walking`, and `GameOver`. There is a third state machine, not pictured, the famous `RedHatBoyStateMachine` that manages the `RedHatBoy` animations. A couple of the states pictured mimic the states in `RedHatBoyStateMachine`, where `Idle` is `Ready` and `Walking` is `Running`, so there is a temptation to move `RedHatBoyStateMachine` into `WalkTheDogStateMachine`. This could work, but remember that `Walk` doesn't have a "jumping" state and so, by doing that, you'll need to start checking a Boolean, and the modeling starts to break down. It's best to accept the similarity because the game is heavily dependent on what RHB is doing, but treat `RedHatBoyStateMachine` as having more fine-grained states. What *does* work is turning `Loading` and `Loaded` into `Option`. Specifically, we'll model our game like so:

```
struct WalkTheDogGame {
    machine: Option<WalkTheDogStateMachine>
}
```

This code isn't meant to be written anywhere yet; it's just here for clarity. There's a big advantage to using `Option` here, and it has to do with the way our `update` function works. For clarity, I'm going to reproduce a section of our game loop here:

```
let mut keystate = KeyState::new();
*g.borrow_mut() = Some(browser::create_raf_closure(move |perf:
f64| {
    process_input(&mut keystate, &mut keyevent_receiver);
    game_loop.accumulated_delta += (perf -
        game_loop.last_frame) as f32;
    while game_loop.accumulated_delta > FRAME_SIZE {
        game.update(&keystate);
        game_loop.accumulated_delta -= FRAME_SIZE;
    }
```

The key part here is the `game.update` line, which performs a mutable borrow on the game object instead of moving it into `update`. This is because once game is owned by `FnMut`, it can't be moved out. Trying to actually leads to this compiler error:

```
error[E0507]: cannot move out of `*game`, as `game` is a
captured variable in an `FnMut` closure
```

Mutable borrows such as this are tricky because they can make it more challenging to navigate the borrow checker as you proceed down the call stack. In this case, it becomes a problem if we try to implement another state machine in the same manner as `RedHatBoyStateMachine`. In our state machine implementation, each `typestate` method consumes the machine and returns a new one. Now, let's imagine that we are modeling the entire game as `enum`, like so:

```
enum WalkTheDogGame {
    Loading,
    Loaded(Walk),
    Walking(Walk),
    GameOver(Walk)
}
```

In order to make this work with the mutable borrow in `update`, we would have to clone the entire game on every state change because the `from` function couldn't take ownership of it. In other words, the closure in our `game.update` function *lends* game to the `update` function. This can't turn around and *give* it to the `from` function – it doesn't own it! Doing so requires cloning the entire game, potentially on every frame!

Modeling the game as holding an optional `WalkTheDogStateMachine` has two advantages:

- We can call `take` on `Option` to get ownership of the state machine.
- The type reflects that the state machine isn't available until the game is loaded.

> **Note**
>
> There are, naturally, many ways to model our game type, and some of them are going to be better than the one we'll choose here. However, before you start trying to do a "simpler" version of this type, let me warn you that I tried several different variations on this solution and ultimately found using `Option` to be the most straightforward choice. Several other implementations either ended with complex borrowing or unnecessary cloning. Be wary, but also be brave. You may find a better way than I did!

Before we dig into the actual implementation, which is fairly long, let's go over the design we're implementing.

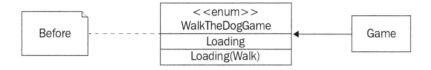

Figure 8.13 – Before

It's pretty simple, but it doesn't do all that we need it to. Now, let's redesign the state machine.

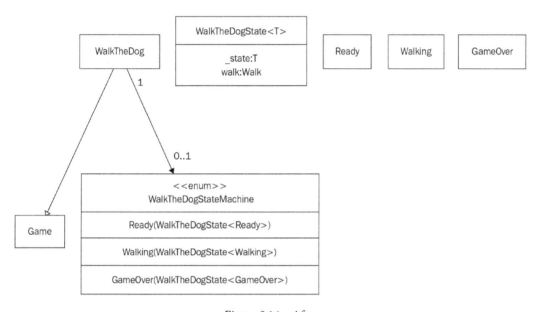

Figure 8.14 – After

Yeah, that's a lot more code, and it doesn't even reflect the details of the implementation, or the From traits we write to make it easy to convert between the enum values and structs. Writing some macros to handle state machine boilerplate is out of the scope of this book, but it's not a bad idea. You might wonder why every state holds its own Walk instance when every single state has it, and that's because we're going to change Walk on the transitions and the individual states don't have easy access to the parent WalkTheDogState container data. However, where possible, we'll move common data out of Walk and into WalkTheDogState.

> **Tip**
> This section has a lot of code, and the snippets tend to only show a few lines at a time so that it's not too much to process. However, as you're following along, you may wish to reorganize the code to be easier to find. For instance, I prefer to work top-down in the `game` module, with constants at the top followed by the "biggest" `struct`, which is `WalkTheDog` in this case, followed by any code it depends on, so that the call stack flows down the page. This is how `https://github.com/PacktPublishing/Game-Development-with-Rust-and-WebAssembly/tree/chapter_8` is organized. You're also welcome to start breaking this up into more files. I won't, to make it easier to explain in book form.

Redesigning to a state machine

In a true refactoring, we would make sure the game was in a running state after each change, but our changes are going to cause cascading compiler errors, meaning we're going to be broken for a while, so this change isn't truly a refactoring but more of a redesign. When you make this kind of change, you should absolutely get to a compiling state as quickly as possible and stay there as long as possible, but while I did that when writing this chapter, I'm not going to put you through all the intermediate steps. We'll move forward as if we know in advance that our design is going to work because we do this time, but don't try this at home. If you're a regular Git user, now is an excellent time to create a branch, just in case.

We'll start by replacing this code in the `game` module:

```
pub enum WalkTheDog {
    Loading,
    Loaded(Walk),
}
```

We'll replace it with the following:

```
pub struct WalkTheDog {
    machine: Option<WalkTheDogStateMachine>,
}
```

This will cause compiler errors all over the place. This is the section where we'll take the shortcut of letting the compiler be broken temporarily while we implement the state machine, if only to make sure this chapter isn't a thousand pages long. So, if you're uncomfortable working with a broken code base for a long time, that's good – just trust that I was *brilliant* and got this all right on the first try. Pretend – it'll be okay.

Since we're psychic and know exactly how this design is going to work out, we can go and push ahead, knowing that eventually, everything will come together without errors. This first change is exactly what we discussed earlier – enum `WalkTheDog` becomes a struct holding its `machine` instance, which is an `optional` field. Currently, `WalkTheDogStateMachine` doesn't exist, so we'll add that next, like so:

```
enum WalkTheDogStateMachine {
    Ready(WalkTheDogState<Ready>),
    Walking(WalkTheDogState<Walking>),
    GameOver(WalkTheDogState<GameOver>),
}
```

When we implement a state machine in Rust, we need enum as a container for states so that `WalkTheDog` doesn't need to be a generic `struct`. We've moved the compiler errors down because there is no `WalkTheDogState` and no states defined. Let's do that next:

```
struct WalkTheDogState<T> {
    _state: T,
    walk: Walk,
}
struct Ready;
struct Walking;
struct GameOver;
```

Right now, the various typestates, `Ready`, `Walking`, and `GameOver`, don't store any data. This will change a little as we go on, but all of the typestates have `Walk` so that they can be saved in the common `WalkTheDogState` struct. Now that we've created the state machine, we need to look at where the old version of `WalkTheDog` was used. The first is in the small `impl` block for `WalkTheDog`, in the old code where we created enum, like so:

```
impl WalkTheDog {
    pub fn new() -> Self {
```

```
            WalkTheDog::Loading {}
    }
}
```

That's not going to work, and it's not compiling, so instead, let's replace it with an empty
`WalkTheDog` instance, as shown here:

```
impl WalkTheDog {
    pub fn new() -> Self {
        WalkTheDog { machine: None }
    }
}
```

This change replaces the old, and not used, enum with `machine` set to `None`. You
can now think of `None` as the `Loading` state, and when a machine is present, you are
`Loaded`. Speaking of loading, the next logical place to make changes is in the `Game`
implementation for `WalkTheDog`. Looking at the `initialize` function that we've been
in so many times, you'll see a compiler error here:

```
#[async_trait(?Send)]
impl Game for WalkTheDog {
    async fn initialize(&self) -> Result<Box<dyn Game>> {
        match self {
            WalkTheDog::Loading => {
```

The `match self` line is not going to work anymore because `self` isn't enum. What
we need to do instead is match `machine`, and if it's `None`, then load the new machine,
and if it's present, then use `Err` in the same way we do now because `initialize` was
somehow called twice. We can start by replacing both halves of the `match` statement, so
the match should start as follows:

```
#[async_trait(?Send)]
impl Game for WalkTheDog {
    async fn initialize(&self) -> Result<Box<dyn Game>> {
        match self.machine {
            None => {
```

Look closely to see that we now match on `self.machine`, and we match against
`None`. Before we dig into the `None` match arm, let's quickly change the match on
`WalkTheDog::Loaded(_)`, as shown here:

```
impl Game for WalkTheDog {
    async fn initialize(&self) -> Result<Box<dyn Game>> {
        match self.machine {
            ...
            Some(_) => Err(anyhow!("Error: Game is already
                initialized!")),
```

This simply changes `WalkTheDog::Loaded` to `Some`, using the same error message.

> **Tip**
>
> In order to get clearer error messages, you can `#[derive(Debug)]` on the
> `WalkTheDog` struct. Doing that has cascading effects because everything it
> depends on also has to `#[derive(Debug)]`, so we won't do that here, but
> it's a good idea, especially if you're running into issues here.

Now that both halves of the match properly match an `Option` type, we need to modify
`initialization` to return the proper type. At the bottom of the `None` branch, you will
want to create a state machine like the one shown here, right before returning the value:

```
impl Game for WalkTheDog {
    async fn initialize(&self) -> Result<Box<dyn Game>> {
        match self.machine {
            None => {
                ...
                let timeline =
                    rightmost(&starting_obstacles);
                let machine = WalkTheDogStateMachine
                    ::Ready(WalkTheDogState {
                    _state: Ready,
                    walk: Walk {
                        boy: rhb,
                        backgrounds: [
                            Image::new(background.clone(),
                                Point { x: 0, y: 0 }),
```

```
                                   Image::new(
                                       background,
                                       Point {
                                               x: background_width,
                                               y: 0,
                                       },
                                   ),
                               ],
                               obstacles: starting_obstacles,
                               obstacle_sheet: sprite_sheet,
                               stone,
                               timeline,
                       },
                   });
                   ...
```

This is very similar to the code before; the construction of `Walk` is unchanged, but it's obscured by all the state machine noise. We are binding the `machine` variable to `WalkTheDogStateMachine::Ready` with the initialized `WalkTheDogState` instance, which, in turn, sets its internal `_state` value to `Ready`, and with the state getting to have `Walk`. It's noisy, and after we get this file back to compiling, we'll do true refactoring to make that line a little cleaner, but put a pin in that for now.

Now, we made it so that `initialize` returns a new `Result<Box<dyn Game>>` a while back, so we'll need to return a new `Game` instance next. So, right after adding `machine`, add the following:

```
impl Game for WalkTheDog {
    async fn initialize(&self) -> Result<Box<dyn Game>> {
        match self.machine {
            None => {
                ...
                Ok(Box::new(WalkTheDog {
                    machine: Some(machine),
                }))
```

> **Note**
>
> Given that `initialize` takes `self` and doesn't really use it, it's debatable
> whether it should be in the `Game` trait. Creating a separate trait, such as
> `Initializer`, will require a lot of modifications and is an exercise for
> the reader.

This takes care of making sure `initialize` returns a game with a machine in the
right state. We have two more big trait methods, `update` and `draw`, to take care of, and
`update` is filled with compiler errors, so let's do that next.

Spreading update into the state machine

The `update` function is filled with compiler errors, is the core of the game's behavior, and
has an additional challenge. Instinctively, you might think you can modify the beginning
of the function like so:

```
impl Game for WalkTheDog {
    ...
    fn update(&mut self, keystate: &KeyState) {
        if let Some(machine) = self.machine {
            ...
```

The `if let Some(machine) = self.machine` line will eventually fail to compile
with the error:

```
error[E0507]: cannot move out of `self.machine.0` which is
behind a mutable reference
   -->src/game.rs:676:32
    |
676 |         if let Some(machine) = self.machine {
    |                    -------      ^^^^^^^^^^^^  help:
consider borrowing here: `&self.machine`
```

Now, you may try, as I did, to fix this by changing the line to `if let Some(machine)
= &mut self.machine`. This will work until you try to implement a transition on
`WalkTheDogState`. Because you have a borrowed machine, you'll also have a borrowed
state when you later match on the state, as with the following example:

```
impl Game for WalkTheDog {
    ...
    fn update(&mut self, keystate: &KeyState)
        if let Some(machine) = &mut self.machine {
```

```
        match machine {
            WalkTheDogStateMachine::Ready(state) => {
```

Here, the `state` value is borrowed, unlike in most other cases where the match arms take ownership of the value, and it's not instantly obvious. It will be if we write a transition from `Ready` to `Walking`. In order to write `state._state.run_right()` and get to `Walking`, your transition will need to look like this in order to compile:

```
impl WalkTheDogState<Ready> {
    fn start_running(&mut self) -> WalkTheDogState<Walking> {
        self.run_right();
        WalkTheDogState {
            _state: Walking,
            walk: self.walk,
        }
    }
}
```

Note that we are transitioning from `&mut WalkTheDogState<Ready>>` to `WalkTheDogState<Walking>`, which is an odd conversion and a hint that this is wrong. What you can't see is that this code won't compile. Returning the new `WalkTheDogState` with `walk` is a move that we cannot do because `state` is borrowed. The `start_running` method doesn't own `state`, so it can't take ownership of `state.walk` and, therefore, can't return the new instance. The workaround for this is to clone the entire `Walk` each time we transition, but there's no need for that inefficiency. We can, instead, take ownership of `machine` all the way back up in the `Game` implementation, through the aptly named `take` function. Instead of using a mutable borrow on the machine, we'll call `take`, as shown here:

```
impl Game for WalkTheDog {
    ...
    fn update(&mut self, keystate: &KeyState) {
        if let Some(machine) = self.machine.take() {
```

This is the same code as earlier, but instead, we call the `take` method on `Option<WalkTheDogStateMachine>`. This replaces the state machine in `self` with `None` and binds the existing `machine` to the variable in `if let Some(machine)`. Now, inside that scope, we have complete ownership of `machine` and can do whatever we like to it, before eventually calling `replace` on the state machine in `self` to move it back in at the end of this function. It's a little awkward, but it gets around mutable borrows. It *also* introduces a potential error in that when control exits the `update` function, `machine` could still be set to `None`, effectively halting the game by accident. In order to prevent that from happening, before we continue updating this function, we'll add `assert` just outside the `if let` statement, as shown here:

```
impl Game for WalkTheDog {
    ...
    fn update(&mut self, keystate: &KeyState) {
        if let Some(machine) = self.machine.take() {
            ...
        }
        assert!(self.machine.is_some());
```

Unfortunately, this is a runtime error, not a compile-time one, but it's going to let us know right away whether we mess up the next section. This `assert` may be overkill, because we are going to dramatically reduce the amount of code inside the `if let` block; in fact, it will be just one line. First, we'll add a call to a non-existent function called `update` on our state machine, as follows:

```
impl Game for WalkTheDog {
    ...
    fn update(&mut self, keystate: &KeyState) {
        if let Some(machine) = self.machine.take() {
            self.machine.replace(machine.update(keystate));

            if keystate.is_pressed("ArrowRight") {
```

Immediately after if let Some(machine), add the self.machine.
replace(machine.update(keystate)) line. All the code below replace in the
if let block is going to become part of various update functions in the implemented
states, so what you'll want to do is either cut and paste that code to some place you can get
it, or just comment it out. Next, we'll create impl on WalkTheDogStateMachine with
this new update method, which will return the new state. An empty version of that will
look like this:

```
impl WalkTheDogStateMachine {
    fn update(self, keystate: &KeyState) -> Self {
    }
}
```

Now, you can call that from the the update method in Game, which looks like this:

```
impl Game for WalkTheDog {
    ...
    fn update(&mut self, keystate: &KeyState) {
        if let Some(machine) = self.machine.take() {
            self.machine.replace(machine.update(keystate));
        }
        assert!(self.machine.is_some());
    }
}
```

The update method in WalkTheDogStateMachine is a little empty, and we should
probably put some code in it. We could call match self in the update, and then write
the behavior for each state in this update function, calling things such as state._
state.walk.boy.run_right(), which would work but it is hideous. Instead, we'll
match on self and then delegate to the individual state types. This will result in a
pretty redundant match statement, as shown here:

```
impl WalkTheDogStateMachine {
    fn update(self, keystate: &KeyState) -> Self {
        match self {
            WalkTheDogStateMachine::Ready(state)
                =>state.update(keystate).into(),
            WalkTheDogStateMachine::Walking(state)
                =>state.update(keystate).into(),
            WalkTheDogStateMachine::GameOver(state)
                =>state.update().into(),
```

```
            }
        }
    }
```

We saw a variation of this pattern before in RedHatBoyStateMachine, where we have
to match on each variant of enum in order to delegate to the state, and unfortunately,
there's not a great way around it. Fortunately, it's small. This little match statement won't
compile because none of the typestates types have an update method. In fact, there
are no implementations for the typestates at all. Let's continue our delegation by creating
placeholder implementations for all three of them, as shown here:

```
impl WalkTheDogState<Ready> {
    fn update(self, keystate: &KeyState) ->
        WalkTheDogState<Ready> {
        self
    }
}
impl WalkTheDogState<Walking> {
    fn update(self, keystate: &KeyState) ->
        WalkTheDogState<Walking> {
        self
    }
}
impl WalkTheDogState<GameOver> {
    fn update(self) -> WalkTheDogState<GameOver> {
        self
    }
}
```

It's worth refreshing our memory on how typestates work. A typestate is a structure
that is generic over a state. So WalkTheDogState<T> is the structure, and we
implement transitions between states by adding methods to implementations of
WalkTheDogState<T>, where T is one of the concrete states. These placeholders all
just return self, so update isn't doing anything yet. Look closely and you'll notice that
GameOver doesn't take KeyState because it won't need it.

The update method on WalkTheDogStateMachine tries to use into to convert each typestate back into enum, but we haven't written those yet. Recalling *Chapter 4, Managing Animations with State Machines*, again, we need to implement From to convert back from the various states to the enum type. These are implemented here:

```
impl From<WalkTheDogState<Ready>> for WalkTheDogStateMachine {
    fn from(state: WalkTheDogState<Ready>) -> Self {
        WalkTheDogStateMachine::Ready(state)
    }
}
impl From<WalkTheDogState<Walking>> for WalkTheDogStateMachine
{
    fn from(state: WalkTheDogState<Walking>) -> Self {
        WalkTheDogStateMachine::Walking(state)
    }
}
impl From<WalkTheDogState<GameOver>> for WalkTheDogStateMachine
{
    fn from(state: WalkTheDogState<GameOver>) -> Self {
        WalkTheDogStateMachine::GameOver(state)
    }
}
```

This is boilerplate just to get things started, but it demonstrates how each of these works. The update method on WalkTheDogStateMachine uses match to get the state value on each variant. Then, the update method is called on the various typestates. Each update method returns the state it transitions into, although right now, they all return self. Finally, back in the update method on WalkTheDogStateMachines, we call into to convert the typestate back into an enum variant.

> **Note**
>
> You might remember that for RedHatBoyStateMachine, we used a transition function and an Event enum to advance the state machine. The new WalkTheDogStateMachine enum has fewer events, so additional complexity isn't necessary.

It's time to think about what each state should actually do. Previously, every one of these states was kind of shoved together in the Game update method – for instance, the following old code:

```
impl Game for WalkTheDog {
    ...
    fn update(&mut self, keystate: &KeyState) {
        if let WalkTheDog::Loaded(walk) = self {
            if keystate.is_pressed("ArrowRight") {
                walk.boy.run_right();
            }

            if keystate.is_pressed("Space") {
                walk.boy.jump();
            }

            if keystate.is_pressed("ArrowDown") {
                walk.boy.slide();
            }
```

In the old system, if the game was Loaded, then boy could run_right if you pressed the ArrowRight button and could jump if you pressed Space. This worked fine, but it's worth noting the following:

- The run_right function does nothing if RHB is already running.

- The jump and slide functions do nothing if RHB isn't running.

We handle this quite well in our RedHatBoyStateMachine, and will continue to do so, but what this reveals is that once RHB starts moving to the right, we don't really care if the player has pushed the **ArrowRight** button, and we don't really care if they push it again. Similarly, if the player hasn't pressed **ArrowRight**, there's no real reason to check whether they pressed **Space** or **ArrowDown**. This all fits well with our new WalkTheDogStateMachine. When the game is Ready, we'll check whether the user has hit ArrowRight and transition the state. Otherwise, we'll just stay in the same state.

We can modify `WalkTheDogState<Ready>` to reflect this new reality. The first change to the function will be to do that check, as shown here:

```
impl WalkTheDogState<Ready> {
    fn update(self, keystate: &KeyState) -> ReadyEndState {
        if keystate.is_pressed("ArrowRight") {
            ReadyEndState::Complete(self.start_running())
        } else {
            ReadyEndState::Continue(self)
        }
    }
}
```

There's one type and one method that doesn't exist, so this code does not compile yet. The transition of `start_running` doesn't exist yet, although we discussed writing something like it. We also don't have the `ReadyEndState` type. Let's address that second one first.

We used this pattern earlier for any `typestate` method that can return more than one state, such as the `update` method on `Jumping` or `Sliding`. We create a new enum that can represent either of the return states. In the case of the `update` method for `WalkTheDogState<Ready>`, the game can either still be `Ready` at the end of an update (`ReadyEndState::Continue`) or be done and transitioning to `Walking` (`ReadyEndState::Complete`).

Let's start by implementing the `From` trait to convert from `ReadyEndState` to `WalkTheDogStateMachine`:

```
enum ReadyEndState {
    Complete(WalkTheDogState<Walking>),
    Continue(WalkTheDogState<Ready>),
}
impl From<ReadyEndState> for WalkTheDogStateMachine {
    fn from(state: ReadyEndState) -> Self {
        match state {
            ReadyEndState::Complete(walking) =>
                walking.into(),
            ReadyEndState::Continue(ready) => ready.into(),
        }
    }
}
```

```
        }
    }
```

This is some boilerplate that you've seen before. We have two states for `ReadyEndState` because there are two states that the `WalkTheDogState<Ready>` update method can end in. In order to get from `ReadyEndState` to `WalkTheDogStateMachine`, we create a `From` trait and match on both variants of `ReadyEndState` and extract their fields from them. Those are both typestates, `WalkTheDogState<Ready>` and `WalkTheDogState<Walking>`, respectively, so we use their `into` methods to convert them into the `WalkTheDogStateMachine` type. Those traits were already written earlier.

The call to `self.start_running` is still not going to work because we haven't written it yet! What happens when the player hits **ArrowRight**? RedHatBoy starts walking! Remember that to transition from one state to another, we write a `typestate` method named after the transition, which looks like so:

```
impl WalkTheDogState<Ready> {
    ...
    fn start_running(mut self) -> WalkTheDogState<Walking> {
        self.run_right();
        WalkTheDogState {
            _state: Walking,
            walk: self.walk,
        }
    }
}
```

Let's refresh our memory on these. Every `state` transition is written as a method on the various typestates – in this case, `WalkTheDogState<Ready>`, where the source state is `self` and the return value is the destination state. Here, we transition from `Ready` to `Walking` by writing a method called `start_running`.

The actual implementation isn't doing much. We start by calling `self.run_right`, which doesn't exist yet, so we have to write it. After sending RHB running, we transition into the `Walking` state by returning a new `WalkTheDogState` instance with `_state` of `Walking`. Take a close look at the function signature for `start_running` and you'll notice it takes `mut state`. This means taking exclusive ownership over `self`, which we can do because we have complete ownership of everything in the state. That is one of the reasons we created `Option<WalkTheDogStateMachine>` originally! However, it's not obvious why we take `mut state` here instead of `state`, in part because `run_right` doesn't exist. When we add our new delegation method, that should become clear, so let's do that right now with the following code:

```
impl WalkTheDogState<Ready> {
    ...
    fn run_right(&mut self) {
        self.walk.boy.run_right();
    }
}
```

This function on `WalkTheDogState<Ready>` calls `run_right` on boy through its walk field. The `run_right` method on boy requires a mutable borrow, and that's why we require a mutable borrow on the previous delegate. It's also why we needed to take `mut state` in the `start_running()` method earlier. You can't mutably borrow something that isn't mutable in the first place.

In order to keep the code clean, we're doing a little more delegation now than we were earlier. This makes our methods smaller and easier to understand, but the trade-off is that the behavior will be spread across multiple places. I think in the end, this will make our code easier to think about, because we won't have to consider too much code at any one time, so the trade-off is worth it. We'll have to be careful that we don't lose track of any of our original code as we break it up into chunks and spread it around.

Re-implementing draw

Now, we've removed all the compiler errors in the original `update` method, in part by removing a large chunk of its functionality, and we can continue by updating the `Walking` state to ensure that it's working, but I believe that's a long time without any meaningful feedback from the game. After all, at this point, the game doesn't compile and doesn't draw. How do we know anything is working? Let's instead take a moment and update the `Game draw` method so that we can actually get the code to compile again and see how it's working.

The `draw` method will start by taking a page from the `update` method and replacing its current implementation with a delegation to `WalkTheDogStateMachine`, as shown here:

```
impl Game for WalkTheDog {
    ...
    fn draw(&self, renderer: &Renderer) {
        renderer.clear(&Rect::new(Point { x: 0, y: 0 },
            600, 600));

        if let Some(machine) = &self.machine {
            machine.draw(renderer);
        }
    }
}
```

There are two things that are a little different from the changes we made to update. The first is that we only borrow `self.machine` because we don't need mutable access. We also still clear the screen at the top of `draw`. That happens on every state change, so there's no reason to not just do it then. Besides, it will help us debug if we make any mistakes, since the screen will turn white.

Let's continue the delegation by adding a `draw` method to `WalkTheDogStateMachine` that can extract the state from each case for drawing, as shown here:

```
impl WalkTheDogStateMachine {
    ...
    fn draw(&self, renderer: &Renderer) {
        match self {
            WalkTheDogStateMachine::Ready(state) =>
                state.draw(renderer),
            WalkTheDogStateMachine::Walking(state) =>
                state.draw(renderer),
            WalkTheDogStateMachine::GameOver(state) =>
                state.draw(renderer),
        }
    }
}
```

This is virtually identical to the `update` method we wrote earlier, except on a borrowed `self` instead of consuming `self`. The rest is just delegations to the various states. Unlike update, every state draws in the exact same way, so we can fill those in with one method, as shown here:

```
impl<T> WalkTheDogState<T> {
    fn draw(&self, renderer: &Renderer) {
        self.walk.draw(renderer);
    }
}
```

Any state will delegate `draw` to `Walk` because the drawing doesn't actually change based on state. We can finally go ahead and re-implement the `draw` method, this time on `Walk`, as shown here:

```
impl Walk {
    fn draw(&self, renderer: &Renderer) {
        self.backgrounds.iter().for_each(|background| {
            background.draw(renderer);
        });
        self.boy.draw(renderer);

        self.obstacles.iter().for_each(|obstacle| {
            obstacle.draw(renderer);
        });
    }
    ...
}
```

This code is not new, but I don't blame you if you forgot it. It's our old `draw` code from *Chapter 6, Creating an Endless Runner*, only with the `walk` variable replaced by `self`. The rest is identical.

At this point, you'll notice something exciting – the code compiles again! But if you look closely at the game, you'll see that it's a little static.

Figure 8.15 – Stand very still....

Red Hat Boy has stopped animating! He doesn't do his little idle animation because we're not calling `update` like we used to; it's almost time to go back to fixing the `update` method.

Refactoring initialize

Before we proceed with restoring functionality, you might remember that I said the creation of `WalkTheDogStateMachine` was "obscured by all the state machine noise." Specifically, it looked like this:

```
let machine = WalkTheDogStateMachine
    ::Ready(WalkTheDogState {
    _state: Ready,
    walk: Walk {
```

To create `WalkTheDogStateMachine` required creating its `Ready` variant and passing a `WalkTheDog` state with its `_state` variable set to `Ready`. In addition to being noisy, it requires you to remember the correct initial state of the state machine. That's what constructors are for!

Let's create a constructor for `WalkTheDogState<Ready>`, as shown here:

```
impl WalkTheDogState<Ready> {
    fn new(walk: Walk) -> WalkTheDogState<Ready> {
        WalkTheDogState {
            _state: Ready,
            walk,
        }
    }
    ...
```

This makes it easier to create a new typestate of `WalkTheDogState<Ready>`; accepting `Walk`, it needs to be valid. Let's also make it easier to create the entire machine, with a smaller constructor:

```
impl WalkTheDogStateMachine {
    fn new(walk: Walk) -> Self {
        WalkTheDogStateMachine
            ::Ready(WalkTheDogState::new(walk))
    }
    ...
```

This constructor creates the entire machine with the right state and passes it `Walk`. Now that we've made these helper methods, we can make the change to the original initialize method, making it a little bit easier to read by using the `WalkTheDogStateMachine` constructor:

```
impl Game for WalkTheDog {
    async fn initialize(&self) -> Result<Box<dyn Game>> {
        match self.machine {
            None => {
                ...
                let machine = WalkTheDogStateMachine
                    ::new(Walk {
                    boy: rhb,
                    ...
```

It's a small change, but makes it both easier to read and safer, too. Doing the right thing, creating `WalkTheDogStateMachine` in the `Ready` state is easy to do, and creating it in the wrong state is not.

Now that we've finished that little digression, we can go back to finishing the update method as planned.

Finishing update

This segment of the *original* `update` function in `Game` reveals what is missing from our current code:

```
impl Game for WalkTheDog {
    ...
    fn update(&mut self, keystate: &KeyState) {
        if let WalkTheDog::Loaded(walk) = self {
            ...
            if keystate.is_pressed("ArrowDown") {
                walk.boy.slide();
            }
            walk.boy.update();
            ...
```

Immediately after all the checks for button presses, we were updating boy. Let's go ahead and add that to our new version of the `update` function in the `WalkTheDogState<Ready>` implementation, like so:

```
impl WalkTheDogState<Ready> {
    fn update(mut self, keystate: &KeyState) ->
        ReadyEndState {
        self.walk.boy.update();
        if keystate.is_pressed("ArrowRight") {
            ReadyEndState::Complete(self.start_running())
        } else {
            ReadyEndState::Continue(self)
        }
    }
    ...
```

There are two changes here, so don't forget to change `update` to accept `mut self` now instead of `self`. It's hiding there in the function signature. Also, we've added a call to `self.walk.boy.update()` to start updating the boy again.

Do that and you'll see RHB idling again, ready to start chasing down his invisible dog. But if you hit the right arrow, RHB freezes, one frame into his running animation. That is not what we want, and intriguingly, there are no errors in the console log because no exceptions are being thrown. It's just that the `Walking` state doesn't do anything in its `update` function. We can restore that code by putting back some of the code we earlier commented out/copied/deleted into the `Walking` state of the game, as shown here:

```
impl WalkTheDogState<Walking> {
    ...
    fn update(mut self, keystate: &KeyState) ->
        WalkTheDogState<Walking> {
        if keystate.is_pressed("Space") {
            self.walk.boy.jump();
        }

        ...

        if self.walk.timeline < TIMELINE_MINIMUM {
            self.walk.generate_next_segment();
        } else {
            self.walk.timeline += walking_speed;
        }
        self
    }
}
```

In `WalkTheDogState<Walking>`, we've modified the `update` method to take a `mut self` and then restored most of the old `Game` update code. Rather than showing the entire method, I've just reproduced the beginning and end of the code snippet and elided the middle; you can safely cut and paste all the original code. There are a few changes to make the code fit its new location. Where the original code would read `walk.boy`, it now reads `self.walk.boy`. I also took the opportunity to rename `velocity`, which is a little vague, to `walking_speed` to clarify that it refers to how fast RHB walks. The final change we've made is taking out the `if keystate.is_pressed("ArrowRight")` code because there's no reason to check for that keypress anymore. Lastly, we return `self` because there's not yet any way to transition out of `WalkTheDogState<Walking>`. If you do this all correctly, you'll find that your code compiles and runs! In fact, as of this moment, all of the behavior is restored, including the problem where we have to refresh to start a new game. How about we finally add a new game button right now, huh?

Start a new Game

If you remember our originally planned behavior, and I don't blame you if you don't, we wanted to draw a new game button on the screen when RHB crashed and fell over. Then, when it's clicked, we want to start a new game. For that to happen, we'll need to do the following:

1. Check whether `RedHatBoyStateMachine` is `KnockedOut`, and if so, transition from `Walking` to `GameOver`.

2. On that transition, draw the new game button.

3. Add an `onclick` handler so that when the button is clicked, we transition back to `Ready` with a new `Walk` instance.

4. On the transition to `Ready`, hide the button and restart the game.

All the code we wrote before was to make that change easier. Let's see whether we were right about that:

1. Transition from Walking to GameOver.

    To transition from Walking to GameOver, we need to return the GameOver state from the WalkTheDogState<Walking> update method, but when should we do that? We'll need to see whether the *boy* is knocked out and then make the change. We don't have that capability yet, so we'll need to create it, and let's work top-down, as we have been this entire chapter. First, we'll change the WalkTheDogState<Walking> update method to check the non-existing method:

```
impl WalkTheDogState<Walking> {

    ...

    fn update(mut self, keystate: &KeyState) ->
        WalkingEndState {

        ...

        if self.walk.timeline < TIMELINE_MINIMUM {
            self.walk.generate_next_segment()
        } else {
            self.walk.timeline += walking_speed;
        }

        if self.walk.knocked_out() {
            WalkingEndState::Complete(self.end_game())
        } else {
            WalkingEndState::Continue(self)
        }
    }
}
```

Now, instead of always returning the `Walking` state, we return `WalkingEndState`, which doesn't exist yet but will mimic the pattern we used in the `update` method on on `WalkTheDogState<Ready>`. When the current state is `knocked_out`, we will return the `Complete` variant holding an instance of the `WalkTheDogState<GameOver>` type. That will be the state returned from the `end_game` transition, which is also not written yet. Otherwise, we'll return `Continue` with the current `WalkTheDogState<Walking>` state as its field. That's two functions that don't exist yet, `knocked_out` and `end_game`, along with a brand-new type. You can create the `WalkingEndState` type and its corresponding `From` trait to convert it into `WalkTheDogStateMachine` right now by following the same pattern we did for `ReadyEndState`. I won't reproduce that code here. We'll proceed from there by getting `knocked_out` working, which is going to be delegated from `Walk` to `RedHatBoyStateMachine` with some delegations in between:

```
impl Walk {
    fn knocked_out(&self) -> bool {
        self.boy.knocked_out()
    }
    ...
}

impl RedHatBoy {
    ...
    fn knocked_out(&self) -> bool {
        self.state_machine.knocked_out()
    }
    ...
}

impl RedHatBoyStateMachine {
    ...
    fn knocked_out(&self) -> bool {
        matches!(self, RedHatBoyStateMachine
            ::KnockedOut(_))
    }
}
```

We could pass `WalkTheDogState` to `RedHatBoyStateMachine` here to get the new state and follow the OO guideline of "tell, don't ask", but sometimes, you just want to check a Boolean. Here, we ask the `Walking` state, which asks `RedHatBoy` and finally `RedHatBoyStateMachine` whether it is knocked out. `RedHatBoyStateMachine` uses the handy `matches!` macro to check `self` against an enum variant, and return whether or not they match. Now that we can check whether Red Hat Boy is knocked out, we have just one compiler error – no method named `end_game` found for struct `WalkTheDogState`.

It's time to implement the `end_game` transition method, which will represent our transition. We can start by implementing the transition to do nothing other than move `walk` from `Walking` to `GameOver`, as shown here:

```
impl WalkTheDogState<Walking> {
    fn end_game(self) -> WalkTheDogState<GameOver> {
        WalkTheDogState {
            _state: GameOver,
            walk: self.walk,
        }
    }
    ...
```

This returns us to a compiled state and means that when RHB crashes and is knocked out, the game is in the `GameOver` state. However, it does nothing, so it's time for *step 2* – draw the new game button.

2. Draw the new game button.

    Many pages ago, I said: "To show our button programmatically, we can just call `browser::draw_ui("<button>New Game</button>")`." But when do we call it? Well, we call it now, right before creating the new state:

```
impl WalkTheDogState<Walking> {
    fn end_game(self) -> WalkTheDogState<GameOver> {
        browser::draw_ui("<button>New Game</button>");
        WalkTheDogState {
            _state: GameOver,
            walk: self.walk,
        }
    }
    ...
```

If you add this one line of code to the transition, you'll see the new game button we wrote way back at the beginning when our RHB crashes into a rock. There's a warning on this line because we don't handle the result of draw_ui, which we'll ignore for the moment.

3. Add the onclick handler to the button.

    In order to add the click handler to the button, we need to get a reference to the element we just drew. We don't have that, as the insert_adjacent_html function doesn't provide it, so we'll need to find the button we just added to the screen so that we can attach an event handler to it. We've used get_element_by_id twice before on document, so it's probably time to write a wrapper function in the browser module, as shown here:

```
pub fn find_html_element_by_id(id: &str) ->
Result<HtmlElement> {
    document()
        .and_then(|doc| {
            doc.get_element_by_id(id)
                .ok_or_else(|| anyhow!("Element with
                    id {} not found", id))
        })
        .and_then(|element| {
            element
                .dyn_into::<HtmlElement>()
                .map_err(|err| anyhow!("Could not cast
                    into HtmlElement {:#?}", err))
        })
}
```

We've made a slight change to the way we've been finding elements in this function. Normally, we want HtmlElement, not a generic Element type, so in this function, we've gone ahead and added a call to dyn_into to make the conversion. Therefore, this function first gets the document, then gets the element, and finally, converts it into the HtmlElement type, all while normalizing the errors with anyhow!.

Now that we have a way to find the element, we can return to the transition in game, find the newly added new game button, and then add a click handler to it, as shown in the following code:

```
impl WalkTheDogState<Walking> {
    fn end_game(self) -> WalkTheDogState<GameOver> {
        let receiver = browser::draw_ui("<button id='new_game'>New Game</button>")
            .and_then(|_unit| browser::
                find_html_element_by_id("new_game"))
            .map(|element| engine::
                add_click_handler(element))
            .unwrap();

        WalkTheDogState {
            _state: GameOver,
            walk: self.walk,
        }
    }
}
```

We've reproduced the entire transition trait here, but there are three changes. The first is that we've added id to the new game button; naturally, that's new_game. Then, we find the element in the document in the and_then block and use map to take that element and pass it to the recently created add_click_handler function. Now, we've got a small problem. We will need receiver to get click messages when they happen, but the add_click_handler function returns Result with UnboundedReceiver. The challenge is that the end_game function doesn't return Result. In *Chapter 9, Testing, Debugging, and Performance*, we'll investigate how to debug this kind of condition, but for now, we'll just grit our teeth and add unwrap.

Now that we have receiver that will get a message whenever the player clicks **New Game**, we need to do something with it. We'll need to check it in the update function for the GameOver state and when we receive the event transition to the Ready state. That's going to mean adding the receiver to the GameOver struct, as follows:

```
struct GameOver {
    new_game_event: UnboundedReceiver<()>,
}
```

This will prompt you to add the use declaration for
`futures::channel::mpsc::UnboundedReceiver`. Now that `GameOver`
struct has the field, we'll need to pass it along in the transition, as shown here:

```
impl WalkTheDogState<Walking> {
    fn end_game(self) -> WalkTheDogState<GameOver> {
        let receiver = browser::draw_ui("<button
            id='new_game'>New Game</button>")
            .and_then(|_unit| browser::
                find_html_element_by_id("new_game"))
            .map(|element| engine::
                add_click_handler(element))
            .unwrap();

        WalkTheDogState {
            _state: GameOver {
                new_game_event: receiver,
            },
            walk: self.walk,
        }
    }
}
```

This is the final change to this method, and it's just adding the field to `GameOver`.
Interestingly it's the first time we've added a field to any of our state structures, but
it's something you're likely to do more of over time as you extend this game. Various
states have data that's unique to them, and they belong in the `state` struct.

It's time to return to the `WalkTheDogState<GameOver>` implementation and
its `update` method, which currently just returns the `GameOver` state, leaving the
game in that state forever. Instead, we'll want to check whether the new game event
has happened (because the button was clicked) and then return the `Ready` state to
start over again. That small bit of code is reproduced here:

```
impl WalkTheDogState<GameOver> {
    fn update(mut self) -> GameOverEndState {
        if self._state.new_game_pressed() {
            GameOverEndState::Complete(self.new_game())
        } else {
            GameOverEndState::Continue(self)
        }
```

```
        }
    }

impl GameOver {
    fn new_game_pressed(&mut self) -> bool {
        matches!(self.new_game_event.try_next(),
        Ok(Some(())))
    }
}
```

In the WalkTheDogState<GameOver> implementation, we check the state to see whether the new game button has been pressed, and if it has, we return the GameOverEndState::Complete variant; otherwise, we return the GameOverEndState::Continue variant. This is the same pattern we've used in every other update method, and you can go ahead and reproduce the GameOverEndState enum and its corresponding From trait to convert the type to a WalkTheDogStateMachine enum. That code is not reproduced here, but remember that if you get stuck, you can find the sample code at https://github.com/PacktPublishing/Game-Development-with-Rust-and-WebAssembly/tree/chapter_8.

In the GameOver implementation, we have the details to check whether new_game_event, corresponding to the player's click, has happened. Calling try_next will return Result immediately, without blocking, or Ok if the channel is still open, regardless of whether anything is in it. Remember that we are running at 60 frames per second and cannot use the blocking calls. Finally, we use the handy matches! macro to check whether the channel was successfully sent a message of unit, or Ok(Some(())). If the event is there, **New Game** has been pressed, and the function returns true.

This code doesn't compile because we don't have a transition written from GameOver to Ready, which is what we'll write in the next step.

4. Restart the game on **New Game**.

Restarting the game will mean doing two things on the new_game transition. The first is hiding the button or "UI," and the second is recreating Walk from scratch. The first is actually easier, so we'll start with that:

```
impl WalkTheDogState<GameOver> {

    ...

    fn new_game(self) -> WalkTheDogState<Ready> {
        browser::hide_ui();
        WalkTheDogState {
            _state: Ready,
            walk: self.walk,
        }
    }
}
```

This is another transition, this time from GameOver to Ready, with the side effect of hiding the UI. It then moves to a new state with the same walk we ended with, which is not quite what we want.

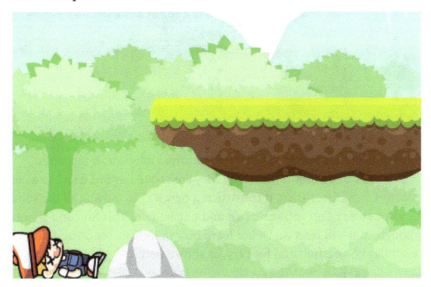

Figure 8.16 – I hit New Game – run, boy, run!

The button is hidden but RHB is still knocked out. Moving from GameOver to Ready means creating a new Walk instance from the old one, so the game starts over. This is a bit of a challenge because we no longer have access to the various images and sprite sheets we used to create Walk and RedHatBoy in the first place. What we'll do is clone those from an existing one, via a constructor function on the Walk implementation. We won't call this clone because that term means an identical copy, whereas this is really a reset. You can see the implementation here:

```
impl Walk {
    fn reset(walk: Self) -> Self {
        let starting_obstacles =
            stone_and_platform(walk.stone.clone(),
                walk.obstacle_sheet.clone(), 0);
        let timeline = rightmost(&starting_obstacles);

        Walk {
            boy: walk.boy,
            backgrounds: walk.backgrounds,
            obstacles: starting_obstacles,
            obstacle_sheet: walk.obstacle_sheet,
            stone: walk.stone,
            timeline,
        }
    }
    ...
}
```

The reset function consumes Walk and returns a new one. It recreates starting_obstacles the same way they are created in initialize, and then recalculates timeline. Then, it constructs a new Walk, moving all the values from Walk except starting_obstacles and timeline. This function is not quite right though, as it will reset Walk but leave boy in its KnockedOut state. We'll need a similar reset function for boy, as shown here:

```
impl RedHatBoy {
    ...
    fn reset(boy: Self) -> Self {
        RedHatBoy::new(
            boy.sprite_sheet,
```

```
                    boy.image,
                    boy.state_machine.context().audio.clone(),
                    boy.state_machine.context().jump_sound.
    clone(),
                )
            }
```

Writing `reset` on `RedHatBoy` is a lot easier than it was on `Walk` because we created a constructor function, `new`, for `RedHatBoy` a long time ago. We should do the same for `Walk`, but that refactoring is up to you. Keep in mind that for this to compile, the `audio` and `jump_sound` fields on `RedHatBoyContext` need to be public.

Now that we have a `reset` function for `RedHatBoy`, we can use it in the `Walk` `reset` function, like so:

```
impl Walk {
    ...
    fn reset(walk: Self) -> Self {
        ...
        Walk {
            boy: RedHatBoy::reset(walk.boy),
            ...
        }
    }
}
```

We also need to call this in the original transition from `GameOver` to `Ready`, as follows:

```
impl WalkTheDogState<GameOver> {
    ...
    fn new_game(self) -> WalkTheDogState<Ready> {
        browser::hide_ui();

        WalkTheDogState {
            _state: Ready,
            walk: Walk::reset(self.walk),
        }
```

```
        }
    }
```

If you do all that, you'll find that when you click the new game button, the game resets and the player is back at the start. You should be able to hit the right arrow key and start walking again. You *should*, but it doesn't work because we haven't accounted for one feature of the UI – the focus.

5. Focus!

It turns out there's one more thing to do when we click the new game button to make the game ready to play again. When the game was started, we set up the canvas to have the focus so that it would receive keyboard input. We did this with the `tabIndex` field in the original HTML. When the player clicks **New Game**, they transfer the focus to the button and then hide the button, which means nothing will get the keyboard events we are listening to. You can see this effect by clicking **New Game** and then clicking the canvas after the button disappears. If you click the canvas, it regains the focus, and you can play the game again.

We can transfer the focus back to the canvas automatically in the `hide_ui` function of the `browser` module. It's debatable whether it belongs here, since you may have cases where you want to hide the UI but not reset the focus, but our game doesn't have that case, so I think we're safe. This change is here:

```rust
pub fn hide_ui() -> Result<()> {
    let ui = find_ui()?;

    if let Some(child) = ui.first_child() {
        ui.remove_child(&child)
            .map(|_removed_child| ())
            .map_err(|err| anyhow!("Failed to remove
                child {:#?}", err))
            .and_then(|_unit| {
                canvas()?
                    .focus()
                    .map_err(|err| anyhow!("Could not
                        set focus to canvas!
                            {:#?}", err))
            })
    } else {
```

```
            Ok(())
        }
    }
```

After the first call to map_err for removing the child, we've added a second and_
then call, which takes unit from the earlier map call, promptly ignores it, and
then requests focus on canvas. The error from the focus call doesn't return an
anyhow! type, so the compiler complains, and we fix that with a map_err call.
The focus function is a JavaScript function we call through web-sys, which is
documented on the MDN (https://mzl.la/30YGOMm).

With that change, you can click **New Game** and start another try. We did it!

Pre-loading

You might notice that the button visibly loads when it shows up on screen – that is to
say that the image and font aren't downloaded to the browser yet, so it doesn't appear
instantaneously. This is standard behavior for web browsers. In order to make sure that
you don't have to wait for an entire page worth of images, fonts, and other assets to load
before you see a page, browsers will load assets lazily. This is so common that your eyes
may not have noticed it when the **New Game** button appeared, but it doesn't look right
in an interactive application. Fortunately, there's a quick way we can fix it. We can tell the
browser to preload the Button.svg and the kenney_future_narrow-webfont.
woff2 assets immediately when the page is loaded so that when the button appears, it's
instantaneous. Open the index.html file and make the changes shown here:

```
<!DOCTYPE html>
<html>
<head>
  <meta charset="UTF-8">
  <title>My Rust + Webpack project!</title>
  <link rel="stylesheet" href="styles.css" type="text/css"
    media=
"screen">
  <link rel="preload" as="image" href="Button.svg">
  <link rel="preload" as="font" href=
  "kenney_future_narrow-webfont.woff2">
</head>
```

The `link` tag with the `preload` attribute will preload assets before rendering the page. You'll want to minimize this behavior generally because you don't want the user to have to wait a very long time with a blank screen, and if you were to make a very large game with many assets, you should probably use a more flexible solution in code with a loading screen. Our game is small right now, so this works perfectly well. With this change, the new game button not only appears but is snappy.

Summary

You can look at the end of this chapter in two ways. The first might be to say, "All that for a button?", and you would have a point. After all, our UI is only one new game button, and while that's true, we actually covered quite a bit. We have integrated the DOM into our app via `web-sys` and have, in turn, adjusted our game to handle it. By utilizing the DOM, we were able to leverage the browser for behavior such as clicks and hovers, without having to detect where within the canvas the mouse was and creating clickable areas. You can now create far more complex UIs using tools such as CSS Grid and Flexbox, so if you are familiar with web development, which you've been doing for this entire book, so you are, you'll be able to make quality UIs for your games. If you're looking for some place to start, try adding a score to this game. You can increment the score in the update, and show it at the end menu, or at the right corner during the game, or both! I look forward to seeing it.

With that, we will move on from new feature development to making sure that our current features work, and work fast. It's now time to start doing some testing and debugging, so we'll dive into that in the next chapter.

Part 3: Testing and Advanced Tricks

While you might have been able to make this small game without testing or debugging, as you develop your own games, you're going to need to prove the games work. You can't do that without writing some tests and debugging some code. You'll want to check performance too, so we'll cover that as well. If you want people to play your game, you'll need a CI/CD pipeline, which is why we'll make one. Finally, we'll look at ways you can expand the game, including interoperating with third-party JavaScript.

In this part, we will cover the following chapters:

9
Testing, Debugging, and Performance

In this book, we've built an entire game using two tools to test our logic – that is, a compiler and our eyes. If the game doesn't compile, it's broken, and if **Red Hat Boy** (**RHB**) doesn't look right, it's broken – simple enough. Fortunately, the compiler provides a lot of tools to make sure we don't make mistakes. Let's be honest, though – it's not enough.

Developing a game can be a long process, especially if you're a hobbyist. When you only have 4 hours to work on it in a given week, they can't all be spent fighting the same bug. To ensure our game works, we need to test it, find mistakes, and make sure it's not too slow. That's what we're going to be doing here.

In this chapter, we will cover the following topics:

- Creating automated tests
- Debugging the game
- Measuring performance with the browser

After completing this chapter, you'll be able to fix the bugs we've written so far and make sure they don't happen again.

Technical requirements

In this chapter, we'll use the Chrome developer tools to debug the code and monitor performance. Other browsers also ship with robust developer tools, but for the screenshots and directions in this chapter, we'll be using Chrome.

The source code for this chapter is available at `https://github.com/PacktPublishing/Game-Development-with-Rust-and-WebAssembly/tree/chapter_9`.

Check out the following video to see the Code in Action: `https://bit.ly/3NKppLk`

Creating automated tests

In an ideal world, every system would have a large amount of testing, both automated and manual, that's done by developers and QA. Some ways to test your game is working correctly involve doing the following:

- Using types to prevent programmer errors
- Playing the game yourself
- Performing automated unit tests
- Performing automated integration tests

So far, we've only used the first two, which is an unfortunately common approach in real-world code. This can be suitable for personal or hobby projects but it isn't robust enough for production applications, particularly those written by a team.

Almost any application can benefit from automated, programmer-written unit tests and as a program becomes even larger, it begins to benefit from integration tests as well. There's not a consistent definition of the differences between these two types of tests as you tend to know them when you see them, but fortunately, we can use the Rust definitions. Rust and Cargo provide two kinds of testing:

- Unit tests via `cargo test`
- Integration tests via `wasm-pack test`

Unit tests tend to be programmer-centric. They are written at the method or function level, with minimal dependencies. You may have a test for every branch of an `if/else` statement, while in the case of a loop, you may have tests for when a list has 0, 1, or many entries. These tests are small and fast and should run in seconds or less. These are my preferred form of testing.

Integration tests tend to look at the app at a higher level. In the case of our code, the integration tests automate the browser and will work based on an event (such as a mouse click) throughout the program. These tests take longer to write, are harder to maintain, and often fail for mysterious reasons. So, why write them? Unit tests typically do not test parts of your application or they may only do so in small doses. This can lead to a situation where your unit tests all pass but the game doesn't work. Most systems will have fewer integration tests than unit tests because of their disadvantages, but they will need them for their benefits.

In Rust, unit tests are written side by side with a module and run with `cargo test`. In our setup, they will run as part of a Rust executable, running directly on the machine. Integration tests are stored in the `tests` directory and only have access to things your crate makes public. They run in the browser – potentially a headless one – with `wasm-pack test`. Unit tests can test internal methods directly, while integration tests must use your crate as a real program would.

> **Tip**
> Ham Vocke has a very detailed article on the *Test Pyramid* that describes one way to organize all of your tests in a system: `https://martinfowler.com/articles/practical-test-pyramid.html`.

Test-driven development

I have a confession to make. I usually write all my code in a test-driven style, where you write a test then make it fail for each step in the development process. Had we followed that process during the development of this book, we'd likely have quite a few tests – perhaps more than 100. In addition, **test-driven development** (**TDD**) exerts a lot of pressure on the design that tends to lead to more loosely coupled code. So, why didn't we do this?

Well, TDD has its downsides, with perhaps the largest being we'd generate a lot more code in the form of tests. We've already written a *lot* of code in this book, so imagine trying to follow along with the tests too – you can see why I felt it was best to leave out the kind of testing I normally write. *Test-Driven Rust* isn't the title of this book after all. However, just because we didn't write tests first doesn't mean we don't want to be sure our code works. That's why, in many cases, we used the type system as the first line of defense against mistakes, such as using the typestate pattern for state transitions. The type system is one of the advantages of using Rust instead of JavaScript for this game.

This isn't to say that automated testing cannot provide value for our program. The Rust ecosystem places a high value on testing, so much so that a testing framework is built into Cargo and is automatically set up for any Rust program. With unit tests, we can test algorithms such as collision detection or our famous state machines. We can make sure that the game still does what we expect, although we can't test whether a game is fun or pretty. For that, you'll have to play the game until you hate it, but a game is a lot more fun if the basics work. We can use tests, along with types, to ensure the code works as expected so that we can turn our focus to whether or not it's fun. To do that, we'll need to set up the test runner and then write some tests that run outside of the browser and inside the browser.

> **Note**
>
> If you're interested in TDD, Kent Beck's book *Test-Driven Development By Example* is still an excellent resource (https://amzn.to/3o1R663). For a web-based approach that uses TypeScript and React, you can take a look at an excellent book called *Build Your Own Spreadsheet* at https://buildyourownspreadsheet.com/.

Getting started

As we mentioned earlier, Rust has built-in capabilities for running tests – both unit and integration. Unfortunately, the template we used way back in *Chapter 1, Hello WebAssembly*, still has an out-of-date setup at the time of writing. If it hasn't been fixed, running cargo test at the command prompt will fail to compile, let alone run the tests. Fortunately, there are not a lot of mistakes. There's just some out-of-date async code for a browser test we won't be using in the automatically generated tests. Those tests are in the tests directory in the app.rs file. This is traditionally where integration tests are put in Cargo projects. We'll change that setup shortly by using unit tests, but first, let's get this to compile by deleting the incorrect async_test setup test. In app.rs, you can delete that function and the #[wasm_bindgen_test(async)] macro above it so that your app.rs file looks like this:

```
use futures::prelude::*;
use wasm_bindgen::JsValue;
use wasm_bindgen_futures::JsFuture;
use wasm_bindgen_test::{wasm_bindgen_test, wasm_bindgen_test_
configure};


wasm_bindgen_test_configure!(run_in_browser);
```

```
// This runs a unit test in native Rust, so it can only use
Rust APIs.
#[test]
fn rust_test() {
    assert_eq!(1, 1);
}

// This runs a unit test in the browser, so it can use browser
APIs.
#[wasm_bindgen_test]
fn web_test() {
    assert_eq!(1, 1);
}
```

> **Note**
> After this book has been published, the template will be fixed and will likely compile. I'm going to assume this, regardless of you changing the code, so that it matches what is here going forward.

Some of the use declarations aren't needed anymore, but they will be short so you can leave them in and ignore the warnings. Now, app.rs contains two tests – one that will run in a JavaScript environment, such as the browser, and one that will run as a native Rust test. Both of these are just examples, where 1 is still equal to 1. To run the native Rust tests, you can run cargo test, as you might be accustomed to. That will run the rust_test function, which is annotated with the test macro. You can run the browser-based tests, which are annotated with the wasm_bindgen_test macro, via the wasm-pack test --headless --chrome command. This will run the web tests using the Chrome browser, in a headless environment. You can also use --firefox, --safari, and --node if you wish, but you must specify what JavaScript environment you'll be running them in. Note that --node isn't going to work since it doesn't have a browser.

We'll start writing tests using the #[test] macro, which runs Rust code in the native environment, just like writing a standard Rust program. The simplest thing to test is a pure function, so let's try that.

Pure functions

Pure functions are functions that, given the same input, will always produce the same output, without side effects such as drawing to the screen or accessing the network. They are analogous to mathematical functions, which just do a calculation, and are by far the easiest types of code to unit test. These tests do not require the browser, so they use the #[test] annotation and run with cargo test.

The current setup runs our only Rust test in the test/app.rs file, which makes it, as far as Cargo is concerned, an integration test. I don't like that and prefer to use the Rust convention of writing unit tests in the file where the code is executed. In this first example, we'll test the intersects function on Rect, which is a pure function that is complicated enough to mess up. We'll add this test to the bottom of engine.rs because that's where Rect is defined, and we'll run it with cargo test. Let's add a test to the bottom of the module for the intersect method on Rect, as shown here:

```rust
#[cfg(test)]
mod tests {
    use super::*;

    #[test]
    fn two_rects_that_intersect_on_the_left() {
        let rect1 = Rect {
            position: Point { x: 10, y: 10 },
            height: 100,
            width: 100,
        };

        let rect2 = Rect {
            position: Point { x: 0, y: 10 },
            height: 100,
            width: 100,
        };

        assert_eq!(rect2.intersects(&rect1), true);
    }
}
```

Much of this is documented in the Rust book at `https://bit.ly/3bNBH3H`, but a little review never hurts anyone. We start by using the `#[cfg(test)]` attribute macro to tell Cargo not to compile and run this code except when we're running tests. Then, we create a `tests` module using the `mod` keyword to isolate our tests from the rest of the code. After that, we import the `engine` code with `use super::*`. Then, we write our test by writing a function, `two_rects_that_intersect_on_the_left`, which is annotated with the `#[test]` macro so that the test runner can pick it up. The rest of this is a pretty standard test. It creates two rectangles, where the second overlaps the first, and then makes sure that the `intersects` function returns `true`. You can run this test with `cargo test`, where you'll see the following output:

```
    Finished test [unoptimized + debuginfo] target(s) in 1.48s
     Running target/debug/deps/rust_webpack_template-
5805000a6d5d52b4


running 1 test
test engine::tests::two_rects_that_intersect_on_the_left ... ok


test result: ok. 1 passed; 0 failed; 0 ignored; 0 measured; 0
filtered out


     Running target/debug/deps/app-ec65f178e238b04b


running 1 test
test rust_test ... ok


test result: ok. 1 passed; 0 failed; 0 ignored; 0 measured; 0
filtered out
```

You'll see two sets of results. The first result references our new test, `two_rects_that_intersect_on_the_left`, which will pass. Then, you will see `rust_test` run, which will also pass. The `rust_test` test is in `tests\app.rs` and was created with the project skeleton. It is run as an integration test because it is in the `tests` directory – this is the Cargo standard. The difference between unit tests and integration tests is that the integration tests are run as a separate crate and use the production code as a separate library. This means they use the code in the same way a user of your crate would, but they cannot call internal or private functions. It's easier to get complete coverage when you're running unit tests with the caveat that they may be less realistic. Our code is not meant to be used as a crate, so we won't be using many integration tests.

Now that we've written our first unit test for our code, we can write a lot more tests for this `intersects` method, including when the following occurs:

- When the rectangles overlap on the top or bottom
- When the rectangles overlap on the right
- When the rectangles *don't* overlap – that is, when the function returns false

We should have a test for every branch in the `intersects` function. We leave these tests as an exercise for you since repeating them would be redundant. As our code base grows, it would be ideal if much of our code could easily be tested like this, but unfortunately, for this game, a lot of it interacts with the browser, so we will have two different ways to test that. The first way is to replace the browser with a stub so that we don't need to run browser-based tests. We'll do that in the next section.

Hiding the Browser module

Way back in *Chapter 3*, *Creating a Game Loop*, we separated browser functions into a `browser` module. We can use this as a **seam** to inject test versions of the browser functions that will run as native Rust code and allow us to write tests.

> **Note**
>
> The term **seam** comes from the book *Working Effectively with Legacy Code*, by Michael Feathers (`https://amzn.to/3kas1Fa`). It's written in C++ and Java but is still the best book on legacy code you can find. A seam is a place where you can insert test behavior to replace real behavior, while an **enabling point** is a point in the code that allows that to happen.

A seam is somewhere we can alter the behavior of the program without altering the code in that place. Look at the following code from the `game` module:

```
impl WalkTheDogState<GameOver> {

    ...

    fn new_game(self) -> WalkTheDogState<Ready> {
        browser::hide_ui();

        WalkTheDogState {
            _state: Ready,
            walk: Walk::reset(self.walk),
        }
```

```
        }
    }
```

We'd like to test that when the game goes from the `GameOver` state to the `Ready` state, the UI is hidden. We can do this with integration tests by checking whether the `div` property that contains the UI is empty after this transition. We may want to do this, but such tests are frequently a little harder to write and maintain. This is especially true as the game grows. Another approach, which we'll use here, is to replace the `browser` module with a version of it that doesn't interact with the browser. The seam is `hide_ui`, which is a behavior we can replace without actually changing the code, while the enabling point is the `use` declaration, which is where we brought in the `browser` module.

We can enable using a test version of the `browser` module with conditional compilation. In the same way that the `#[cfg(test)]` macro only includes the `test` module when compiling in test mode, we can import different versions of the `browser` module with `cfg` directives, as shown here:

```
#[cfg(test)]
mod test_browser;
#[cfg(test)]
use test_browser as browser;

#[cfg(not(test))]
use crate::browser;
```

The preceding code can be found at the top of the `game` module, where we were previously importing `crate::browser`. Here, we can use the `mod` keyword to bring the contents of the `test_browser` module in from the `src/game/test_browser.rs` file, but only when we're running a `test` build. Then, we can use `test_browser as browser` to make the functions available via `browser::` calls – again, only in test builds – in the same way as we call the `browser` production code. Finally, we can add the `#[cfg(not(test))]` annotation to `use crate::browser` to prevent the real `browser` code from being imported into the test.

> **Note**
> I first saw this technique on Klausi's Weblog at `https://bit.ly/3ENxhWQ`, but it is fairly common in Rust code.

If you do this and run `cargo test`, you'll see a lot of errors, such as `cannot find function 'fetch_json' in module 'browser'`, because even though we're importing a test module, we haven't filled it in with any code. In this situation, it's a good idea to follow the compiler errors, which will point out that there's no file yet in `src/game/test_browser.rs`. It will also list the functions that are used in the `game` module but aren't defined in our `test_browser.rs` file. To get past this, you can create the `test_browser.rs` file and bring in the bare minimum that's needed to get back to compiling, as shown here:

```
use anyhow::{anyhow, Result};
use wasm_bindgen::JsValue;
use web_sys::HtmlElement;

pub fn draw_ui(html: &str) -> Result<()> {
    Ok(())
}

pub fn hide_ui() -> Result<()> {
    Ok(())
}

pub fn find_html_element_by_id(id: &str) -> Result<HtmlElement>
{
    Err(anyhow!("Not implemented yet!"))
}

pub async fn fetch_json(json_path: &str) -> Result<JsValue> {
    Err(anyhow!("Not implemented yet!"))
}
```

As you can see, only four functions are used in game that have been defined in browser, and we've filled in just enough to compile. To use this for testing, we're going to need to place simulated implementations with some sort of state they keep track of. The other thing you may notice is that JsValue and HtmlElement are both being used in this code since they won't work when you run Rust native tests. They require a browser runtime, so to continue along this path, we'll eventually need to make test versions of HtmlElement and JsValue or create wrapper types for them, potentially in the engine module. Let's leave those as is for now, though, and try to write our first test using the standard Rust testing framework. We'll want to test the state machine change I mentioned previously by setting up the game in the GameOver state and transitioning to the Running state, then checking that the UI was hidden. The *beginning* of that test looks as follows:

```
#[cfg(test)]
mod tests {
    use super::*;
    use futures::channel::mpsc::unbounded;
    use std::collections::HashMap;
    use web_sys::{AudioBuffer, AudioBufferOptions};


    fn test_transition_from_game_over_to_new_game() {
        let (_, receiver) = unbounded();
        let image = HtmlImageElement::new().unwrap();
        let audio = Audio::new().unwrap();
        let options = AudioBufferOptions::new(1, 3000.0);
        let sound = Sound {
            buffer: AudioBuffer::new(&options).unwrap(),
        };
        let rhb = RedHatBoy::new(
            Sheet {
                frames: HashMap::new(),
            },
            image.clone(),
            audio,
            sound,
        );
        let sprite_sheet = SpriteSheet::new(
            Sheet {
```

```
            frames: HashMap::new(),
        },
        image.clone(),
    );
    let walk = Walk {
        boy: rhb,
        backgrounds: [
            Image::new(image.clone(), Point { x: 0, y:
                0 }),
            Image::new(image.clone(), Point { x: 0, y:
                0 }),
        ],
        obstacles: vec![],
        obstacle_sheet: Rc::new(sprite_sheet),
        stone: image.clone(),
        timeline: 0,
    };
    let state = WalkTheDogState {
        _state: GameOver {
            new_game_event: receiver,
        },
        walk: walk,
    };
    }
}
```

Oh boy – that's a lot of code to test a few lines of Rust, and it's not even a complete test yet. It's just setting up the game in the state that we need it to be in *before* we transition into a Ready state. A lot is being revealed about our design, specifically that it's what I may call *naïve'*. It's very hard to construct objects, and while the game, engine, and browser modules are separate, they are still pretty tightly coupled. It works but it in a fashion that only solves the problem in front of us. That's completely acceptable – we had specific goals to build a small endless runner and we did it, but this also means that if we wanted to start extending our game engine so that it's more flexible, we would need to make further changes. I tend to view software design more like sculpting than constructing. You start with a big block of code and chip away at it until it looks like what you want, rather than a blueprint that you follow to get the perfect house.

Some of the aspects of our design that this test is revealing are as follows:

- It's not easy to create new `Walk` structures.

- The `game` module is far more coupled to `web-sys` and `wasm-bindgen` than we thought.

We made the intentional choice not to try and create perfect abstractions early in the project. This is one of the reasons we didn't write this code in a test-driven style initially. TDD would have strongly pushed in the direction of further abstraction and layering, which would have hidden the game code we're trying to learn here. As an example, instead of using `HtmlImageElement` or `AudioBuffer`, we may have written wrappers or abstractions around those objects (we already have an `Image` struct), which is probably better for growing our project in the medium to long term but is harder to understand in the short term.

This is a long-winded way of saying that this code is now hard to write isolated unit tests for because we didn't build it with them in mind. If you were able to run this test, you would see the following:

```
thread 'game::tests::test_transition_from_game_over_to_new_
game' panicked at 'cannot call wasm-bindgen imported functions
on non-wasm targets', /Users/eric/.cargo/registry/src/
github.com-1ecc6299db9ec823/web-sys-0.3.52/src/features/gen_
HtmlImageElement.rs:4:1
```

It turns out that even though we replaced the production `browser` with `test_browser`, we're still trying to call browser code. I have already pointed out `HtmlElement` and `JsValue`, but this test also includes `AudioBuffer` and `AudioBufferOptions`. As is, this code doesn't compile without more feature flags being enabled and changes being made to `engine`. It's just too tightly coupled to the browser still.

The act of trying to use this code in a test harness demonstrated the power of coupling, and it is often useful to take legacy code and get it into a harness to identify these dependency problems and break them. Unfortunately, this is a time-consuming process that we are not going to continue using in this section, although it may appear on my blog at `paytonrules.com` at some point. Instead, we'll test this code via a test that runs in the browser.

Browser tests

At the beginning of this chapter, I mentioned that there were **unit tests** and **browser tests**. The distinction is that while browser tests may test the same behavior as a unit test, they automate the desired behavior in a headless browser. This makes the test more realistic, but also slower and more prone to breaking for flaky reasons. I prefer my systems to have a large base of unit tests and a smaller number of more integrated tests to make sure everything is all wired together correctly, but we can't always get what we want.

Instead, we'll get what we need – verification of the behavior – by skipping dependency-breaking techniques for legacy code and writing a test that runs in the browser. We'll remove the code that added the `test_browser` module, as well as the `test_browser` file itself. We'll keep the test we wrote previously and make two changes for it to compile, as follows:

1. Add `AudioBufferOptions` to the list of `web-sys` features in `Cargo.toml`.

2. In the `engine` module, make the `buffer` field on the `Sound` struct public so that we can create `Sound` directly in this test.

These two changes will get the code compiling, but it won't make it run in the tests yet. For that, we need to make a couple of changes. First, we need to change the `#[test]` macro to `#[wasm_bindgen_test]`. Then, we need to add two statements to our `test` module, as shown here:

```
#[cfg(test)]
mod tests {
    use super::*;
    use futures::channel::mpsc::unbounded;
    use std::collections::HashMap;
    use web_sys::{AudioBuffer, AudioBufferOptions};

    use wasm_bindgen_test::wasm_bindgen_test;

    wasm_bindgen_test::wasm_bindgen_test_configure!
        (run_in_browser);
    #[wasm_bindgen_test]
    fn test_transition_from_game_over_to_new_game() {
        ...
```

The first line to add is use `wasm_bindgen_test::wasm_bindgen_test` so that the macro is present. The second is `wasm_bindgen_test::wasm_bindgen_test_configure!(run_in_browser);`. This directive tells the test runner to run in the browser so that the code can interact with the DOM, similar to how the application does. This test won't run in `cargo test`, so you'll need to use the `wasm-pack test --headless -chrome` command. This will run the web tests in a headless version of the Chrome browser. When you run them, you should see the following output:

```
running 1 test
```

```
test rust_webpack_template::game::tests::test_transition_from_
game_over_to_new_game … ok
```

Now, we have a test that's running and passing, but the only problem is that we don't have any assertions. We've written an "arrange" step but we haven't checked the results. The point of this test was to make sure that the UI was hidden when the state transition happened, so we'll need to update the test to check that. We can do this by adding the action and assertion steps, as shown here:

```
#[wasm_bindgen_test]
fn test_transition_from_game_over_to_new_game() {
    ...
    let document = browser::document().unwrap();
    document
        .body()
        .unwrap()
        .insert_adjacent_html("afterbegin", "<div
            id='ui'></div>")
        .unwrap();
    browser::draw_ui("<p>This is the UI</p>").unwrap();
    let state = WalkTheDogState {
        _state: GameOver {
            new_game_event: receiver,
        },
        walk: walk,
    };
```

```
    state.new_game();

    let ui =
        browser::find_html_element_by_id("ui").unwrap();
    assert_eq!(ui.child_element_count(), 0);
}
```

Here, we start the test by inserting the `div` property, along with the `ui` ID, into the document – after all, that is in `index.html` in the game. Then, `browser::draw_ui` draws the UI to the browser, even though the browser is running headlessly, so we don't see it. We continue by creating `WalkTheDogState` in the `GameOver` state; on the next line, we have it transition to `Ready` via the `state.new_game()` method. Finally, we check that the UI was cleared by finding the `div` property and checking its `child_element_count`. If it's 0, the code is right, and this test will pass. If you run this test, you'll see that this test *does* pass, so you will probably want to comment out the `let next_state: WalkTheDogState<Ready> = state.new_game()` line and run it again just to make sure it fails when the transition happens.

This is still a very long test but at least it's working. The test can be cleaned up by creating some factory methods in the various modules so that structs are easier to create. You'll notice that the test is full of `unwrap` calls. This is because, in a test, I want things to crash right away if they aren't as expected. Unfortunately, browser-based tests with the `wasm_bindgen_test` macro do not let you return a `Result` for readability as standard Rust tests do yet. This is another reason you should try and make your tests run as native Rust tests.

Async tests

One of the biggest challenges of testing web applications, whether they're Wasm or traditional JavaScript ones, is code that occurs **asynchronously**. In the case of our code, that's anything that runs in an `async` block or function. Imagine calling a function in an `async` test and then immediately trying to verify it worked. By definition, you can't, because it's running asynchronously and may not have finished yet. Fortunately, `wasm_bindgen_test` handles this rather easily by making the test's functions `async` themselves.

Let's look at a simpler example and try to write a test for the `load_json` function in the `browser` module:

```
#[cfg(test)]
mod tests {
    use super::*;
    use wasm_bindgen_test::wasm_bindgen_test;

    wasm_bindgen_test::wasm_bindgen_test_configure!
        (run_in_browser);

    #[wasm_bindgen_test]
    async fn test_error_loading_json() {
        let json = fetch_json("not_there.json").await;

        assert_eq!(json.is_err(), true);
    }
}
```

This can be found in the `browser` module. Here, we start with the boilerplate to set up a `tests` module, import both `browser` and `wasm_bindgen_test`, and configure the test to run in the browser. The test itself is only two lines. Try to load a JSON file that doesn't exist and report an error. The key difference between this is that the test is `async`, which allows us to use `await` in the test and write the assertion without adding any "wait for" logic. This is great, but there are a couple of things to keep in mind:

- If `fetch_json` can hang, this test will hang.
- This test will try to load a file. Ideally, we don't want to do this in a unit test.

This test will run and pass. We could test all of the `browser` functions this way, accepting that the `browser` module's tests will use the filesystem as needed. That's probably what I would do if I was handed this system in a professional environment. You could work very hard to stub out the actual browser on these tests, but to do so would remove its ability to prevent defects. After all, if you remove the browser from the `browser` module, then how do you know you got the code right?

If I was given this code and asked to maintain it, I would likely adopt the following strategies:

- Curse the name of the jerk who wrote it without tests (me!).

- Write tests for code as I need to change it. If it doesn't change, don't bother. Go ahead and use browser automation, as we did previously.

- Over time, move more code that depends on `wasm-bindgen` and `web-sys` into the `browser` module so that `engine` and `game` can stub it out.

- Write as many tests as possible as Rust-native tests, and then make the browser-based unit tests native whenever possible.

As for integration tests, I doubt I would write any integration tests in the Cargo sense. For Cargo libraries, all the integration tests are written in the `tests` directory and compiled as a separate package. This is a great idea when you're writing a library that's going to be consumed by other people, but we are writing an application and aren't providing an API. The integration tests I would write would be any tests that use the real browser, but those are integration tests in the sense that they are integrated with the web browser, not that they run as Rust integration tests.

However, we can't just rely on adding tests to make sure our code works. Sometimes, we just have to debug it. Let's dig into that next.

Debugging the game

To debug a traditional program, be it in Java, C#, or C++, we must set breakpoints and step through the code. In JavaScript, we can type the word `debugger` to set a breakpoint, but although WebAssembly runs in the browser, it isn't JavaScript. So, how do we debug it?

There's a lot of conflicting information about debugging with WebAssembly. How do you debug WebAssembly? Well, according to the official Rust WebAssembly documentation, it's simple – you can't!

Unfortunately, the debugging story for WebAssembly is still immature.
On most Unix systems, DWARF is used to encode the information that a
debugger needs to provide source-level inspection of a running program.
There is an alternative format that encodes similar information on
Windows. Currently, there is no equivalent for WebAssembly. Therefore,
debuggers currently provide limited utility, and we end up stepping through
raw WebAssembly instructions emitted by the compiler, rather than the
Rust source text we authored.

– `https://rustwasm.github.io/docs/book/reference/`
`debugging.html`

So, there you have it – no debugging, section over. That was easy.

But it's not that simple. Of course, you can debug your application – you just can't use your browser's developer tools to step through the Rust code in a debugger. The technology isn't there yet. But that doesn't mean we don't debug; it just means we'll take more of an old-school approach to debugging.

Earlier, I mentioned that when I write code, I typically write a lot of tests. I also typically don't use a debugger very often. If we break our code into smaller chunks that can be easily exercised by tests, a debugger is rarely required. That said, we didn't do that for this project, so we'll need a way to debug existing code. We'll start by logging, then getting stack traces, and finally using linters to prevent bugs before they happen.

> **Note**
>
> The reality is not as cut and dry as the Rust Wasm site would state. Chrome developer tools have added support for the **DWARF** debugging format to the browser, as detailed here: `https://developer.chrome.com/blog/wasm-debugging-2020/`. This standard format, whose specification can be found at `https://dwarfstd.org/`, unfortunately is not supported by `wasm-bindgen` at the time of writing. You can see progress on this issue here: `https://github.com/rustwasm/wasm-bindgen/issues/2389`. By the time you read this book, the debugging tools may be modernized in Rust Wasm, as well as in browsers outside of Chrome, but for the time being, we must use more traditional tools such as `println!` and logging.

Log versus error versus panic

If you've been following along and got confused at some point, then you've probably used the `log!` macro we wrote in *Chapter 3, Creating a Game Loop*, to see what was going on. If you have been doing that, congratulations! You've been debugging the same way I did when I wrote the code originally. Print line debugging is still standard in many languages and it's pretty much the only form of debugging that's guaranteed to work anywhere. If you haven't done that, then it looks like this:

```
impl WalkTheDogStateMachine {
    fn update(self, keystate: &KeyState) -> Self {
        log!("Keystate is {:#?}", keystate);
        match self {
            WalkTheDogStateMachine::Ready(state) =>
                state.update(keystate),
            WalkTheDogStateMachine::Walking(state) =>
                state.update(keystate),
            WalkTheDogStateMachine::GameOver(state) =>
                state.update(),
        }
    }
}
```

In the preceding example, we are logging `KeyState` on every tick through the `update` function. This isn't a great log because it's going to show an empty `KeyState` 60 times a second, but it's good enough for our purposes. However, there's one flaw in this log: `KeyState` doesn't implement the `Debug` trait. You can add it by adding the `derive(Debug)` annotation to the `KeyState` struct, like so:

```
#[derive(Debug)]
pub struct KeyState {
    pressed_keys: HashMap<String, web_sys::KeyboardEvent>,
}
```

When you add this, the console will log all your key state changes, which will be useful if your keyboard input is broken:

```
151 Keystate is KeyState {
        pressed_keys: {},
    }
 6 Keystate is KeyState {
        pressed_keys: {
            "ArrowRight": KeyboardEvent {
                obj: UiEvent {
                    obj: Event {
                        obj: Object {
                            obj: JsValue(KeyboardEvent),
                        },
                    },
                },
            },
        }
    }
   Keystate is KeyState {
        pressed_keys: {
            "Space": KeyboardEvent {
                obj: UiEvent {
                    obj: Event {
                        obj: Object {
                            obj: JsValue(KeyboardEvent),
                        },
                    },
                },
            },
            "ArrowRight": KeyboardEvent {
                obj: UiEvent {
                    obj: Event {
                        obj: Object {
                            obj: JsValue(KeyboardEvent),
                        },
                    },
                },
            },
        },
    }
 2 Keystate is KeyState {
        pressed_keys: {
            "Space": KeyboardEvent {
                obj: UiEvent {
                    obj: Event {
                        obj: Object {
                            obj: JsValue(KeyboardEvent),
                        },
                    },
                },
            },
        },
    },
```

Figure 9.1 – Logging KeyState

In general, any `pub struct` should use `#[derive(Debug)]`, but this isn't the default option since it could make compile times long on large projects. When in doubt, go ahead and use `#[derive(Debug)]` and log the information. Now, maybe `log!` isn't noticeable enough for you, and you want the text to be bright, obvious, and red. For that, you'll need to use `console.error` in JavaScript and write a macro such as the `log` macro, which we already have in the `browser` module. This macro looks like this:

```
macro_rules! error {
    ( $( $t:tt )* ) => {
        web_sys::console::error_1(&format!( $( $t )*
            ).into());
    }
}
```

This is the same as the `log` macro but uses the `error` function on the `console` object. There are two advantages to the `error` function. The first is that it's red, while the other is that it also will show you the stack trace. Here's an example of `error` being called when the player is knocked out in Chrome:

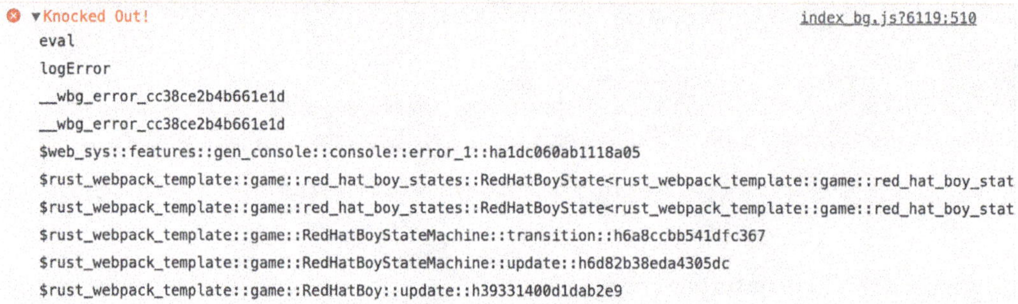

```
⊗ ▼Knocked Out!                                                              index_bg.js?6119:510
    eval
    logError
    __wbg_error_cc38ce2b4b661e1d
    __wbg_error_cc38ce2b4b661e1d
    $web_sys::features::gen_console::console::error_1::ha1dc060ab1118a05
    $rust_webpack_template::game::red_hat_boy_states::RedHatBoyState<rust_webpack_template::game::red_hat_boy_stat
    $rust_webpack_template::game::red_hat_boy_states::RedHatBoyState<rust_webpack_template::game::red_hat_boy_stat
    $rust_webpack_template::game::RedHatBoyStateMachine::transition::h6a8ccbb541dfc367
    $rust_webpack_template::game::RedHatBoyStateMachine::update::h6d82b38eda4305dc
    $rust_webpack_template::game::RedHatBoy::update::h39331400d1dab2e9
```

Figure 9.2 – Error log

It's not the most readable stack trace in the world, but after seeing a few lines of the `console::error_1` function, you can see that this log was called from `WalkTheDogState<Walking>::end_game`. This log is really for true errors, as opposed to just informational logging, and this stack trace may not show up clearly in all browsers. You'll also want to be cautious with leaving this log in the production code as you may not want to expose this much information to a curious player. We'll want to make sure it's not in the production deployment, which we'll create in *Chapter 10, Continuous Deployment*.

Finally, if you want to make sure the program stops when an error occurs, we'll want to go ahead and use the panic! macro. Some errors are recoverable but many are not, and we don't want our program to limp along in a broken state. In *Chapter 1*, *Hello WebAssembly*, we included the console-error-panic-hook crate so that if the program were to panic, we'd get a stack trace. Let's replace calling error! with calling panic! and see the difference:

```
⊗ ▶ panicked at 'Knocked Out!', src/game.rs:829:13                                          index_bg.js?6119:318

  Stack:

  Error
      at eval (webpack-internal:///./pkg/index_bg.js:415:15)
      at logError (webpack-internal:///./pkg/index_bg.js:309:18)
      at Module.__wbg_new_693216e109162396 (webpack-internal:///./pkg/index_bg.js:414:48)
      at __wbg_new_693216e109162396 (http://localhost:8080/index.js:77:98)
      at console_error_panic_hook::Error::new::h9e80bb47a581a2b0 (http://localhost:8080/b1b0701…module.wasm:wasm-function[333
  4]:0xdc4fe)
      at console_error_panic_hook::hook_impl::h888f0c3ffa3b9a0f (http://localhost:8080/b1b0701…module.wasm:wasm-function[388]:0
  x76a6c)
      at console_error_panic_hook::hook::h40b6708eaa09f954 (http://localhost:8080/b1b0701…module.wasm:wasm-function[3703]:0xe15
  20)
      at core::ops::function::Fn::call::h6db1003a4d38595c (http://localhost:8080/b1b0701…module.wasm:wasm-function[3070]:0xd83d
  1)
      at std::panicking::rust_panic_with_hook::h606d7c7f7a423b98 (http://localhost:8080/b1b0701…module.wasm:wasm-function[838]:
  0x9797a)
      at std::panicking::begin_panic_handler::{{closure}}::h9b985a293aac4ce1 (http://localhost:8080/b1b0701…module.wasm:wasm-fu
  nction[1477]:0xb3791)

⊗ ▶ Uncaught RuntimeError: unreachable                                                       b1b0701…module.wasm:0xe8565
      at __rust_start_pani(http://localhost:8080/     n:0xe8565)
      at rust_panic (b1b0701…module.wasm:0xe5829)          b1b0701f883780de69e.module.wasm
      at std::panicking::rust_panic_with_hook::h606d7c7f7a423b98 (b1b0701…module.wasm:0x979a1)
      at std::panicking::begin_panic_handler::{{closure}}::h9b985a293aac4ce1 (b1b0701…module.wasm:0xb3791)
      at std::sys_common::backtrace::__rust_end_short_backtrace::ha03abef02a8b70fd (b1b0701…module.wasm:0xe815c)
      at rust_begin_unwind (b1b0701…module.wasm:0xdf27b)
      at core::panicking::panic_fmt::h6314b5c91abe7349 (b1b0701…module.wasm:0xe3d97)
      at
  rust_webpack_template::game::red_hat_boy_states::RedHatBoyState<rust_webpack_template::game::red_hat_boy_states::Falling>::kno
  ck_out::ha651747d2994ca8b (b1b0701…module.wasm:0xbc973)
```

Figure 9.3 – Panic log

Here, you can see it looks a little different, but the information is mostly the same. There is one thing at the very top where it says src/game.rs:829, which tells you exactly where panic was called. In general, you will probably prefer to use panic compared to error if you need to have the error in your production code because that kind of error should be rare and fail fast. The error function is more useful during debugging, so you'll end up removing those.

There's another kind of error that we've been ignoring at times, and that's the warnings and errors that are given to you by the compiler and linter. We can use the Rust ecosystem's tools to detect mistakes before we ever run the program. Let's look into that now.

Linting and Clippy

One of the features that makes the Rust compiler great is that it has a linter built into it, in addition to the warnings and errors it already provides. If you're unfamiliar, a linter is a static code analysis tool that typically finds style errors and, potentially, logic errors above and beyond what the compiler can find. The term comes from the lint you find on clothing, so you can think of using a linter like rubbing a lint brush on your code. We've been getting some warnings from the compiler that we've been ignoring for a while now, most of which look like this:

```
warning: unused `std::result::Result` that must be used
  --> src/game.rs:241:9
     |
241 |            browser::hide_ui();
     |            ^^^^^^^^^^^^^^^^^^^^
     |
     = note: `#[warn(unused_must_use)]` on by default
     = note: this `Result` may be an `Err` variant, which should
be handled
```

These are all cases where an error could occur, but we probably don't want to crash if it does, so panicking or calling unwrap isn't an option. Propagating the Result type is an option, but I don't think we want to prevent moving from one state to another if there's a small browser issue. So, instead, we'll use the error case to log here. You can see it at https://bit.ly/3q1936N in the sample source code. Let's modify the code so that we log any errors:

```
impl WalkTheDogState<GameOver> {
    ...
    fn new_game(self) -> WalkTheDogState<Ready> {
        if let Err(err) = browser::hide_ui() {
            error!("Error hiding the browser {:#?}", err);
        }

        WalkTheDogState {
            _state: Ready,
            walk: Walk::reset(self.walk),
        }
```

```
        }
    }
```

Here, we have changed the `browser::hide_ui()` line to `if let Err(err) = browser::hide_ui()` and we log if an error occurs. We can see what that error log will look like by forcing `hide_ui` to return an error for a moment:

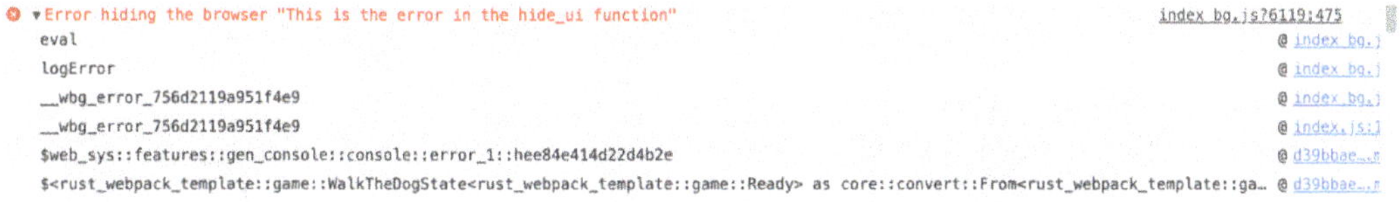

```
⊗ ▾Error hiding the browser "This is the error in the hide_ui function"                          index_bg.js?6119:475
    eval                                                                                          @ index_bg.j
    logError                                                                                      @ index_bg.j
    __wbg_error_756d2119a951f4e9                                                                  @ index_bg.j
    __wbg_error_756d2119a951f4e9                                                                  @ index.js:j
    $web_sys::features::gen_console::console::error_1::hee84e414d22d4b2e                          @ d39bbae...m
    $<rust_webpack_template::game::WalkTheDogState<rust_webpack_template::game::Ready> as core::convert::From<rust_webpack_template::ga... @ d39bbae...m
```

Figure 9.4 – A fake error

The stack trace is cut off in book form, but you can see that we got an error log with `Error hiding the browser` and then `This is the error in the hide_ui function`, which is the error message I forced into `hide_ui`. The stack trace also shows `game::Ready`, which would show you that you were transitioning into the `Ready` state if you had infinite room to show the entire message.

Every single warning that's being generated should be dealt with. Most of the warnings are the same kind – that is, `Result` types where the `Err` variant is ignored. These can be removed by handling the `Err` case with a log or by calling `panic` if the game should truly crash at this time. For the most part, I've used the `if let` pattern but if `request_animation_frame` fails, then I just use `unwrap`. I don't see how the game could work if that's failing.

There is one more warning we've been ignoring that we should address, as shown here:

```
warning: associated function is never used: `draw_rect`
  --> src/engine.rs:106:12
   |
106 |        pub fn draw_rect(&self, bounding_box: &Rect) {
   |               ^^^^^^^^^
   |
   = note: `#[warn(dead_code)]` on by default
```

This warning is a little unique because we used this function *for debugging*. You may not want to draw rectangles in your game, but it's essential for debugging collision boxes, as we did in *Chapter 5, Collision Detection*, so we'll want it to be available. To keep it around, let's annotate it with the `allow` keyword, like so:

```
impl Renderer {
    ...


    #[allow(dead_code)]
    pub fn draw_rect(&self, bounding_box: &Rect) {
```

This should leave the compilation error-free, but there's one more tool we can use to see whether our code could be improved. If you've spent much time in the Rust ecosystem, then you've probably heard of **Clippy**, a popular Rust linter that will catch common Rust mistakes and improve your code, above and beyond those found by the compiler's defaults. It's installed as a Cargo component, so it's not added to your `Cargo.toml` file but to the current system itself. Installation is simple, and you may have done it at some point and forgotten about it, but if you haven't, it's one shell command:

```
rustup component add clippy
```

Once you've installed Clippy, you can run `cargo clippy` and see all the other ways we wrote bad Rust code.

> **Note**
>
> When the code is great, I wrote it and you followed along. When it's bad, we did it together. I don't make the rules.

When I run `cargo clippy`, I get 17 warnings, but your number could be different, depending on when you run it. I'm not going to go through each one, but let's highlight one error:

```
warning: writing `&Vec<_>` instead of `&[_]` involves one more
reference and cannot be used with non-Vec-based slices.
  --> src/game.rs:945:29
   |
945 |  fn rightmost(obstacle_list: &Vec<Box<dyn Obstacle>>) ->
i16 {
   |                              ^^^^^^^^^^^^^^^^^^^^^^^^^ help:
change this to: `&[Box<dyn Obstacle>]`
```

```
          |
      = note: `#[warn(clippy::ptr_arg)]` on by default
      = help: for further information visit https://rust-lang.
  github.io/rust-clippy/master/index.html#ptr_arg
```

The `rightmost` function in the `game` module can be made to use one less reference and be made more flexible. `help` here is great because it tells me exactly what to do to fix it. So, let's change the `rightmost` function signature so that it looks as follows:

```
fn rightmost(obstacle_list: &[Box<dyn Obstacle>]) -> i16 {
```

This doesn't fix any bugs but it does remove a Clippy warning and makes the method more flexible.

It's very common for Clippy to inform you of better idioms you could be using. One Clippy warning I wanted to highlight looks like this:

```
warning: match expression looks like `matches!` macro
    --> src/game.rs:533:9
      |
533 | /              match self {
534 | |                  RedHatBoyStateMachine::KnockedOut(_) =>
true,
535 | |                  _ => false,
536 | |              }
    | |_____^ help: try this: `matches!(self,
RedHatBoyStateMachine::KnockedOut(_))`
      |
      = help: for further information visit https://rust-lang.
github.io/rust-clippy/master/index.html#match_like_matches_
macro
```

I had this error occur quite a bit in earlier versions of the code. I wasn't aware that the `matches!` macro existed before I ran Clippy, but what it does is handle the exact case where you need to check whether an `enum` is a specific case you're looking for. That's why the code now uses what Clippy suggests, which is in `impl RedHatBoyStateMachine`:

```
impl RedHatBoyStateMachine {
    ...
    fn knocked_out(&self) -> bool {
```

```
matches!(self, RedHatBoyStateMachine::KnockedOut(_))
}
```

> **Tip**
>
> Many editors make it very easy to enable Clippy as part of syntax checking so that you don't need to run it explicitly. If you can enable it, you should do so.

Many of the other errors are about overusing `clone` and using `into` when it isn't necessary. I highly recommend going through the code and fixing those, taking another moment to understand why they were flagged. In *Chapter 10, Continuous Deployment*, we'll add Clippy to our build process so that we don't have to keep putting up with these errors.

At this point, the code has been tested (a little) and we've handled every compiler error and warning we can think of. It's safe to say that the game works, but is it fast enough? The next thing to check is its performance. So, let's do that now.

Measuring performance with a browser

The first step in debugging performance is answering the question, *Do you have a performance problem?* Too many developers, especially game developers, worry too early about performance and introduce complex code for a performance gain that just isn't there.

For example, do you know why so much of this code uses `i16` and `f16`? Well, when I was going back to school a few years ago, I took a game optimization class in C++, where our final project needed to optimize a particle system. The biggest performance gains were to convert 32-bit integers into 16-bit integers. As my professor would say, "*We got to the moon on 16-bit!*" So, when I was writing this code, I internalized the lesson and made the variables 16-bit unless they were being sent to JavaScript, where everything is 32-bit anyway. Well, allow me to quote directly from the WebAssembly specification (found at `https://webassembly.github.io/spec/core/syntax/types.html`):

Number types classify numeric values.

The i32 and i64 types classify 32- and 64-bit integers, respectively. Integers are not inherently signed or unsigned; their interpretation is determined by individual operations.

The f32 and f64 types classify 32- and 64-bit floating-point data, respectively. They correspond to the respective binary floating-point representations, also known as single and double precision, as defined by the IEEE 754-2019 standard (Section 3.3).

It turns out that WebAssembly doesn't support a 16-bit numeric value, so all of the optimization to `i16` is pointless. It's not harming anything and it's not worth going back to change it, but it reinforces the first rule of optimization: **measure first**. With that in mind, let's investigate two different ways to measure the performance of our game.

Frame rate counter

There are two ways our game can perform poorly: by using too much memory and slowing the frame rate. The second of those is far more important, especially for a small game like this, so we'll want to start looking at frame rate first. If the frame rate consistently lags, our game loop will account for it as best it can, but the game will look jittery and respond poorly. So, we need to know the current frame rate, and the best way to do that is to draw it on the screen.

We'll start by adding a function, `draw_text`, that will draw arbitrary text on the screen. This is debug text, so similarly to the `draw_rect` function, we'll need to disable the warning that says the code is unused. Writing text is a function of `Renderer` in the `engine` module, as shown here:

```
impl Renderer {
    ...
    #[allow(dead_code)]
    pub fn draw_text(&self, text: &str, location: &Point) ->
Result<()> {
        self.context.set_font("16pt serif");
        self.context
            .fill_text(text, location.x.into(),
                location.y.into())
            .map_err(|err| anyhow!("Error filling text
                {:#?}", err))?;
        Ok(())
    }
}
```

We've hardcoded the font here because this is for debugging purposes only, so it's not worth customizing. Now, we need to add a frame rate calculator to the game loop, which is in the `start` method of `GameLoop` in the `engine` module. You can refresh your memory on how it works by reviewing *Chapter 3, Creating a Game Loop*. The frame rate can be calculated by taking the difference between the last two frames, dividing by 1,000, to get from milliseconds to seconds, and calculating its inverse (which is 1 divided by the number). This is simple but it will lead to the frame rate fluctuating wildly on screen and won't show very useful information. What we can do instead is update the frame rate every second so that we can get a fairly stable indicator of performance on screen.

Let's add that code to the `engine` module. We'll start with a standalone function that will calculate the frame rate every second in the `start` method, as shown here:

```rust
unsafe fn draw_frame_rate(renderer: &Renderer, frame_time: f64)
{
    static mut FRAMES_COUNTED: i32 = 0;
    static mut TOTAL_FRAME_TIME: f64 = 0.0;
    static mut FRAME_RATE: i32 = 0;

    FRAMES_COUNTED += 1;
    TOTAL_FRAME_TIME += frame_time;

    if TOTAL_FRAME_TIME > 1000.0 {
        FRAME_RATE = FRAMES_COUNTED;
        TOTAL_FRAME_TIME = 0.0;
        FRAMES_COUNTED = 0;
    }

    if let Err(err) = renderer.draw_text(
        &format!("Frame Rate {}", FRAME_RATE),
        &Point { x: 400, y: 100 },
    ) {
        error!("Could not draw text {:#?}", err);
    }
}
```

Oh no – it's an `unsafe` function! It's the first one in this book, and probably the last. We're using an `unsafe` function here because of the `static mut` variables – that is, FRAMES_COUNTED, TOTAL_FRAME_TIME, and FRAME_RATE – which are not safe in a multithreaded environment. We know that this function won't be called in a multithreaded way, and we also know that if it was called, it would just show a weird frame rate value. It's not something I generally recommend, but in this case, we don't want to pollute `GameLoop` or the `engine` module with those values or put them in thread-safe types. After all, we wouldn't want to have our frame rate calculator take too long because of a bunch of `Mutex` lock calls. So, we'll accept that this debugging function is `unsafe`, shiver in fear for a moment, and move on.

The function starts by setting up the initial FRAMES_COUNTED, TOTAL_FRAME_TIME, and FRAME_RATE values. On each call to `draw_frame_rate`, we update TOTAL_FRAME_TIME and the number of FRAMES_COUNTED. When TOTAL_FRAME_TIME has passed `1000`, this means that 1 second has elapsed, since TOTAL_FRAME_TIME is in milliseconds. We can set FRAME_RATE to the number of FRAMES_COUNTED because that's the literal **frames per second** (**FPS**) and then reset both counters. After calculating the frame count, we draw it with the new `draw_text` function we just created. This function is going to be called last on every frame, which is important because if it isn't, we would draw the game right over the top of the frame rate. If we didn't draw the frame rate on every frame, we also wouldn't see it except for brief flickers on the screen, which is hardly suitable for debugging.

Now, let's add the call to `GameLoop`, in the `start` function, as shown here:

```
impl GameLoop {
    pub async fn start(game: impl Game + 'static) -> Result<()>
    {
        ...
        *g.borrow_mut() = Some(browser::create_raf_closure
            (move |perf: f64| {
            process_input(&mut keystate, &mut
                keyevent_receiver);

            let frame_time = perf - game_loop.last_frame;
            game_loop.accumulated_delta += frame_time as
                f32;
            while game_loop.accumulated_delta > FRAME_SIZE {
                game.update(&keystate);
                game_loop.accumulated_delta -= FRAME_SIZE;
```

```
        }
        game_loop.last_frame = perf;
        game.draw(&renderer);

        if cfg!(debug_assertions) {
            unsafe {
                draw_frame_rate(&renderer, frame_time);
            }
        }
        ...
```

The `game_loop.accumlated_delta` line has changed slightly, pulling the calculation for the length of the frame into a temporary variable, `frame_time`. Then, after drawing, we check whether we are in debug/development mode through the check for `if cfg!(debug_assertions)`. This will ensure that this doesn't show up in the deployed code. If we are in debug mode, we call `draw_frame_rate` inside an `unsafe` block. We send that function `renderer` and `frame_time`, which we just pulled into a temporary variable. Adding this code gives us a clear measurement of the frame rate on the screen:

Frame Rate 60

Figure 9.5 – Showing the frame rate

On my machine, the frame rate is a steady `60`, with an occasional blip that isn't consistent. That's great unless you're writing a chapter on debugging performance issues. Then, you may have a problem.

Fortunately, in early drafts, there was one time when the frame rate dropped, and that was when the RHB crashed into a rock. When the **New Game** button showed up, the frame rate suddenly dropped, albeit briefly, and did so every time. We can "restore" that defect by removing the preloaded button and font from `index.html`. In other words, we must delete the highlighted code in `index.html`:

```
<!DOCTYPE html>
<html>
<head>
  <meta charset="UTF-8">
  <title>My Rust + Webpack project!</title>
```

```
<link rel="stylesheet" href="styles.css" type="text/css"
    media=
"screen">
<link rel="preload" as="image" href="Button.svg">
<link rel="preload" as="font" href=
"kenney_future_narrow-webfont.woff2">
</head>
```

If you delete the preloaded assets, you should see the see frame rate dip briefly. Displaying the frame rate is a great way to make sure that you, as a developer, see performance issues right away. If the frame rate dips, then you've got a problem, just like we have when we don't preload the assets. Sometimes, we need more than just a frame rate counter. So, let's leave the preload code deleted and see the performance problem in the browser debugger.

Browser debugger

Every modern browser has developer tools, but I'll be using Chrome for this section as it's the one most popular with developers. In general, they all look similar to each other. To get performance information, I must start the game and open the developer tools in Chrome. Then, I must right-click and click **Inspect**, though there are plenty of other ways to open the tools. From there, I must click the **Performance** tab and start recording. Then, I must run RHB into a rock and stop recording. Since I know I've got a specific spot with a performance dip, I want to get to it as quickly as possible to hide any noise in the debugger from other code. After I do that, I will see a graph, like this:

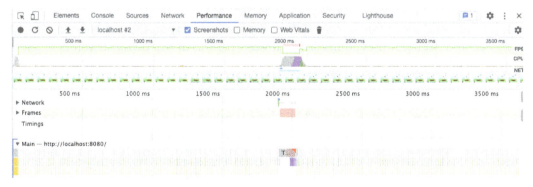

Figure 9.6 – The Performance tab

That's a lot of noise, but you can see that the graph changes. There's a pink blob on the **Frames** row, which shows that something happened there. I can select the section that looks like a hill with my cursor and drag it to zoom in on it. Now, I will see the following screen:

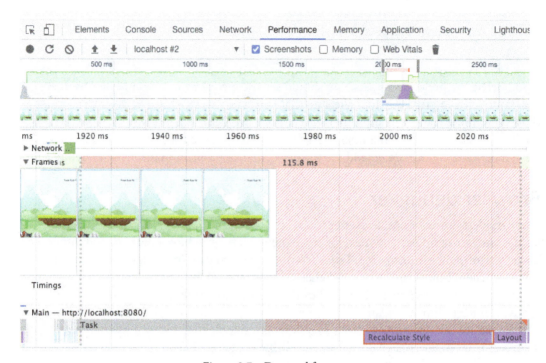

Figure 9.7 – Dropped frames

Here, you can see that one frame was **115.8 ms**. I opened the **Frames** section (see how the gray arrow next to **Frames** points down) to see what was drawn on those frames – our poor knocked-out RHB. A frame that's 115.8 ms is way too long, and if you hover your mouse over that, it will show you **dropped frames**. Beneath the **Frames** section, there's the **Main** section, which shows what the application was doing. I've highlighted **Recalculate Style** here, which is taking **33.55 ms** according to the **ToolTip** window, which shows up after I roll my mouse over it.

Recalculate Style is something the browser has to do when you add things to the DOM, such as a button. How did we write our UI again? We added buttons to the UI. By inserting the button into the document, we force the styles to be recalculated and redo the layout; since we didn't preload the elements, it's slower than one of our frames in a game. To speed this up, we can restore the three preloaded lines we deleted from the `index.html` file, which should speed up recalculating the layout. If you do that and remeasure your performance, you'll see something like this:

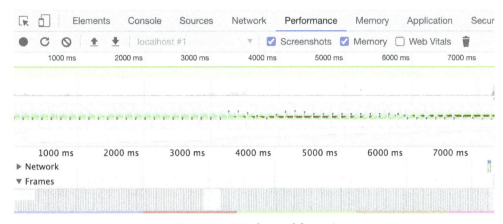

Figure 9.8 – No dropped frames!

Was this worth worrying about? Possibly – it is noticeable to see the button load, but it's not worth extending this chapter to fix it. You know how to fix it, and you know how to find the issue in the **Performance** tab, and that's what's important for now. Anyway, we have another question to answer: how much memory is this game taking up?

Checking memory

When I was writing this game, I would frequently leave it running all day in the background, only to have my browser become very unresponsive as it started taking up all my computer's memory. I began to suspect that the game had a memory leak, so I started investigating. You may think it's impossible to have a memory leak in Rust due to its guarantees, and it is harder, but remember that a lot of our code talks to JavaScript, where we don't necessarily have the same guarantees. Fortunately, we can check this with the same tools we have been using to test performance.

Go ahead and clear the performance data by clicking **no sign** in the top-left corner. Then, start another recording and play for a little while. This time, don't try to die right away; go ahead and let the game play for a bit. Then, stop recording and look at the performance data again, this time ensuring you click the **Memory** button. Now, you can a look at the results, which may look like this:

Figure 9.9 – Memory profiling

Can you see that blue wave at the bottom of the screen, which shows **HEAP** in the bottom right-hand corner? This shows that our memory grows and then is periodically reclaimed. This may not be ideal as we'd like memory to be constant, but we aren't trying to control things to that degree at this time. Chrome, and most browsers, run their garbage collectors in separate threads so that they won't affect performance as much as you may think. It would be worth experimenting and creating a memory budget in the application and keeping all the allocations in that budget, but that's outside the scope of this book. Fortunately, the memory is reclaimed and it doesn't look like the game is growing uncontrollably.

After further investigation, it turned out that the problem with my browser was caused by my company's bug tracker, which uses far more memory than this little game! If you're seeing performance issues, make sure you account for other tabs, browser extensions, and anything else that might be slowing down your computer outside of the game.

Summary

This chapter was a little different than the previous ones because, in many ways, our game is complete! But of course, it's not perfect, which is why we spent some time looking at ways we can investigate defects and bullet-proof the code base.

We dug into automated testing, writing unit tests for our transitions, and writing integration tests that run in the browser. We now have logging for any unforeseen errors and stack traces if the code crashes, both of which are necessary diagnostics for debugging challenging errors. Then, we used the linter and Clippy to clean up our code and remove subtle issues that the compiler can't catch. Finally, we investigated performance issues in the browser and found that we had none!

In the next chapter, we'll get those tests into a CI/CD setup and even deploy them to production. What are we waiting for? Let's ship this thing!

10
Continuous Deployment

The traditional way to publish a game is to create a main copy of the build and ship it off to manufacturing. This was frequently referred to as **going gold** inside and outside of the gaming industry, and it still is if you're making a AAA game that's being shipped to consoles and sold in stores. The process is time-consuming and extremely expensive; fortunately, we don't have to do it! Walk the Dog is a web-based game that we need to ship to a website. Since we're deploying to the web, we can use all the best practices of the web, including continuous deployment, where we'll deploy a build whenever we want directly from source control.

In this chapter, we'll cover the following topics:

- Creating a **Continuous Integration/Continuous Delivery (CI/CD)** pipeline
- Deploying test and production builds

When this chapter is complete, you'll be able to ship your game to the web! How else will you become rich and famous?

Technical requirements

In addition to a GitHub account, you'll need a Netlify account. Both of these have significant free tiers, so if cost becomes a problem, then congratulations! Your game took off! You'll also need to be familiar with Git. You don't need to be an expert, but you'll need to be able to create repositories and push them to GitHub. If Git is completely new to you, then the GitHub *Getting Started* guide is a good place to start: `https://docs.github.com/en/get-started`. The sample code for this chapter is available at `https://github.com/PacktPublishing/Game-Development-with-Rust-and-WebAssembly/tree/chapter_10`.

Check out the following video to see the Code in Action: `https://bit.ly/3DsfDsA`

Creating a CI/CD pipeline

When you run `npm run build` locally, a release build is put inside the `dist` directory. Theoretically, you could take that directory and copy it to a server somewhere to deploy your application. This will work provided that the server knows about the `wasm` MIME type, but copying to a directory manually is a very old-fashioned way of deploying software. Nowadays, we automate the build and deploy on a server, along with additional code that's been checked into source control. It's significantly more complicated than the old-fashioned way, so why is it better?

The practice of automating the build this way is often referred to as CD and its definition is pretty big. Take a look at the following quote from `https://continuousdelivery.com`:

> *Continuous Delivery is the ability to get changes of all types—including new features, configuration changes, bug fixes, and experiments—into production, or into the hands of users, safely and quickly in a sustainable way.*

You might read this and think that yes, copying from your machine's `dist` directory onto a server is exactly that, but it isn't. A few issues can happen when deploying manually. We've listed a few of them here:

- The documentation could be wrong or lacking, meaning only one person knows how to deploy.

- The deployed code might not be the same as the code in source control.

- Deployments might only work based on a local configuration, such as the version of `rustc` that exists on an individual's machine.

There are many more reasons why you don't simply want to run npm run build locally and then copy/paste to a server. But when a team is small, it's very tempting to say, "I'll worry about it later." Instead of listening to that little voice, let's try to think about the qualities of a deployment that are safe and quick, as the definition says. We can start with the opposite of some of the preceding bullet points. If those are reasons why a manual deployment does not qualify as CD, then a process that does qualify would be able to do the following:

- Automate the process so that it is repeatable by everybody on the team.

- Always deploy from source control.

- Declare the configuration in source control, so it's never incorrect.

There's a lot more to a proper CD process than the preceding list. In fact, a "perfect" CD is often more of a goal to be reached than an end state that you hit. Since we're a one-person band, we won't be hitting every single bullet point from the *Continuous Delivery* book (https://amzn.to/32bf9bt), but we will be making a **pipeline** that builds code, runs tests, and then deploys a test build on **pull requests** (**PRs**). Then, on merges to main, it will deploy to a production site. For this, we'll use two technologies: GitHub Actions and Netlify.

> **Note**
> CI refers to the practice of frequently merging code into the primary branch (main in Git parlance) and running all the tests to ensure the code still works. CI/CD is a shorthand for combining the practices of integration and delivery, although it's a bit redundant since CD includes CI.

GitHub Actions is a relatively new technology from GitHub. It is used for running tasks when branches are pushed to GitHub. It's well suited for running CI/CD because it's built right into the source control that we're already using and has a pretty good free tier. If you decide to use a different tool, such as Travis CI or GitLab CI/CD, you can use this implementation to guide how you would use those other tools. At this point, the similarities outnumber the differences.

After running CI on GitHub Actions, we'll deploy to Netlify. You might be wondering why we're using Netlify if our stated goal is to reduce the number of new technologies, and that's because, while we can deploy directly to GitHub Pages, that won't support creating test builds. In my opinion, an important part of a good CD process is the ability to create production-like builds that can be experimented on and tried out. Netlify will provide that out of the box. If your team has grown from beyond one person, you'll be able to try out the game as part of the process of reviewing code in a PR. Also, Netlify is set up to work with Wasm out of the box, so that's handy.

> **Note**
>
> In GitHub parlance, a PR is a branch that you wish to merge into the `main` branch. You create a PR and ask for a review. This branch can run other checks before being allowed to be merged into the `main` branch. Other tools, such as GitLab, call these **merge requests** (**MRs**). I tend to stick to the term PR because it's what I'm used to.

Our pipeline will be fairly simple. On every push to a PR branch, we'll check out the code, build and run the tests, then push to Netlify. If the build is a branch build, you'll get a temporary URL to test out that build. If the push is to `main`, then it will deploy a release build. In the future, you might want a little more rigor around production deployments, such as tagging releases with release notes, but this should be fine to get us started.

The first step is to make sure the build machine is using the same version of Rust that we're using locally. The `rustup` tool allows you to install multiple versions of the Rust compiler along with multiple toolchains, and you'll want to make sure that everybody on the team and along with CI is using the same version of Rust. Fortunately, `rustup` provides several different ways of doing this. We'll use the `toolchain` file, which is a file that specifies the toolchain for the current project. In addition to ensuring any machine that builds this crate will use the same version of Rust, it also documents the version of Rust used for development. Every Rust project should have one.

> **Note**
>
> At the time of writing this chapter, I discovered that I had made a mistake in the first draft of *Chapter 1, Hello WebAssembly*. I hadn't documented the Rust version being used or ensured that the `wasm32-unknown-unknown` toolchain was installed. These are the exact kinds of errors that come up when you try to set up a CI build, because you've forgotten all of those early assumptions, and it's also one of the reasons why it's important to have a CI build. Sadly, you can always forget documentation, but the build machine can't lie. This is why I frequently set up CI at the beginning of a project.

The `toolchain` file is named `rust-toolchain.toml` and is kept at the root directory of the crate. We can create one that looks like this:

```
[toolchain]
channel = "1.57.0"
targets = ["wasm32-unknown-unknown"]
```

The preceding toolchain says we'll use version `1.57.0` of the Rust compiler and the `wasm32-unknown-unknown` target, so we can be sure we'll be able to compile to WebAssembly. Now that we've ensured the version of Rust we're using, we can start setting up a CI/CD pipeline in GitHub Actions. You're welcome to try newer versions, but this has been tested using `1.57.0`.

GitHub Actions

Like many other CI/CD tools, GitHub Actions is defined by the configuration files in your source repository. When you create the first configuration file, called a *workflow* in Actions, it will get picked up by GitHub, which will then start a *runner*. You can see the output in the **Actions** tab of a GitHub repository. The following screenshot shows what the tab looked like for me while writing this chapter:

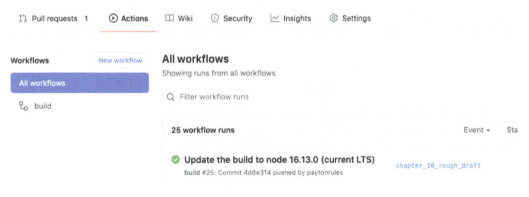

Figure 10.1 – A green build

This is an example workflow being run on GitHub, where I have updated the version of the deployment to use the LTS version of Node.js. It's a little unfortunate that you have to go to the **Actions** tab to see the result of your *workflows*, but I suppose marketing won out. It's also a little confusing to hear the terms *workflow* and *pipeline* thrown around. A *workflow* is a specific GitHub Actions term referring to a series of steps run on its infrastructure via the configuration we'll build next. A *pipeline* is a CD term referring to a series of steps that are needed to deploy software. So, I can have a pipeline made up of one or more workflows if I'm running it on GitHub Actions and using their terminology. This pipeline will be made up of one workflow, so you can use them interchangeably.

To begin building our pipeline, we'll need to ensure we have a GitHub repository for Walk the Dog. You probably already have one, but if you don't, you have two options to choose from:

- Create a new repository from your existing code.
- Fork the example code.

You can do either, although it would be a shame if the code you've been writing all along didn't exist in a repository somewhere. If you do fork from my repository, then make sure you fork from the *Chapter 9, Testing, Debugging, and Performance* sample code at `https://github.com/PacktPublishing/Game-Development-with-Rust-and-WebAssembly/tree/chapter_9`. Otherwise, all the work will be done for you. In either case, from now on, I'll assume you have your code in a repository on GitHub.

> **Tip**
>
> If, at any point, you find yourself confused, you can cross-check the GitHub Actions documentation at `https://docs.github.com/en/actions/learn-github-actions/understanding-github-actions`. We'll try to keep the workflow simple, so you won't need to be an Actions expert.

We can start setting up a workflow with a kind of "Hello World" for GitHub Actions. The workflow will simply check the code, and it should turn green almost immediately after pushing. Create a file, named `.github/workflows/build.yml`, and add the following YAML to it:

```
on: [push]

name: build
```

```
jobs:
  build:
    runs-on: ubuntu-latest
    steps:
      - uses: actions/checkout@v2
```

YAML (Yet Another Markup Language) is the markup language of many CI/CD pipelines. If you've never seen it before, note that it is whitespace sensitive. This means that, sometimes, if you copy/paste it from one file to another or from a book into code, it might not be syntactically correct. Here, I'm using two spaces per tab, which is the standard format, and YAML does not allow tab characters.

> **Tip**
>
> YAML is mostly self-explanatory, and it's also not the important takeaway from this chapter. So, if there's some YAML syntax that confuses you, it's probably not worth worrying about. But just in case, there is a pretty good YAML cheat sheet at `https://quickref.me/yaml`.

For the most part, you can read YAML as a list of key-value pairs. This workflow starts with the `on` key, which will run this workflow on every `push` event. It's an array, so you can set up workflows for multiple events, but we won't be doing that. The next key, `name`, gives the workflow a name. Then, we add the `jobs` key, which will only have one job. In our case, it is `build`. We specify that our job runs on `ubuntu-latest` with the `runs-on` key. Then, finally, we define its list of steps. This job currently only has one step, `uses: actions/checkout@v2`, and that's worth explaining in more depth. Each step can either be a shell command or—you guessed it—an *action*. You can create your own actions, but most actions are created by the GitHub community; they can be found in the GitHub Actions marketplace.

You might be able to guess that `actions/checkout@v2` checks the code, and you'd be right. But you're probably wondering where that comes from and how you were supposed to know about it. That's where the Actions marketplace comes in, which can be found at `https://github.com/marketplace?type=actions`:

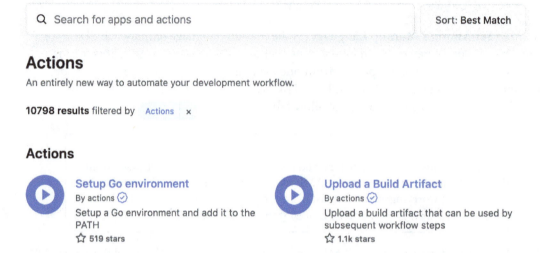

Figure 10.2 – The Actions marketplace

Your workflow is made up of a series of steps run in order, most of which are found on the GitHub marketplace. Don't let the name "marketplace" fool you; actions don't cost money. They are open source projects and free as in beer. Let's dig into the first action we'll be using (`https://github.com/marketplace/actions/checkout`):

Figure 10.3 – Checkout

The checkout action can be found in almost every single workflow since it's pretty hard to do anything without checking out the code first. If you browse this page, you'll see there's full-featured documentation for the action, along with a big green button that says **Use latest version**. If you click on that button, a small snippet is presented to you, showing you how to integrate the action into your workflow:

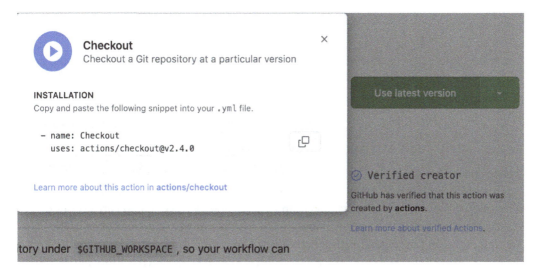

Figure 10.4 – Copy and paste me!

These actions are the building blocks of workflows. Setting up a CI/CD pipeline in GitHub Actions means searching through the marketplace, adding actions to your workflow, and reading the docs. This is significantly easier than the tangle of Bash scripts that I used in the past, although don't worry, you can call trusty Bash scripts, too.

> **Note**
>
> I want to emphasize that this isn't meant to be taken as an endorsement of GitHub Actions over any other CI/CD solution. Nowadays, there are so many great tools for this kind of work that it's hard to recommend one tool over another. I've used Travis CI and GitLab CI/CD quite a bit over the years, and they are also great. That said, GitHub Actions is also pretty great.

If you commit this change and push it to a branch (do *not* use `main` for now) inside your repository, you can check the **Actions** tab to see the workflow running successfully, as shown in the following screenshot:

Figure 10.5 – Checking out the code

We have checked out the code, and now we need to build it on the GitHub *runner*. A runner is just a fancy name for a machine. To build Rust on your local machine, you need the `rustup` program with an installed compiler and toolchain. We could run a series of shell scripts; however, instead, we will look to see whether any Rust actions exist in the marketplace. I won't hold you in suspense—there's an entire library of Rust-related actions to be found at `https://actions-rs.github.io/`. It's a great collection, and it will make it easier to create our build. We'll add steps to do the following:

- Install a toolchain (`https://actions-rs.github.io/#toolchain`).
- Install wasm-pack (`https://actions-rs.github.io/#install`).
- Run Clippy (`https://actions-rs.github.io/#clippy-check`).

The preceding links will take you to the official documentation for each of the actions, all of which have been created and maintained by Nikita Kuznetsov (`https://svartalf.info/`). Since each action is specified in YAML, it can use any keys it likes. Potentially, this means a lot of flags and configurations to document, but we'll stick to the straightforward flag.

So, what are we waiting for? Let's add the step required to install a toolchain, as shown here:

```
steps:
  - uses: actions/checkout@v2
  - uses: actions-rs/toolchain@v1
    with:
      toolchain: 1.57.0
```

```
        target: wasm32-unknown-unknown
        override: true
        components: clippy
```

I've left the checkout step in the sample for reference, but the code we've added starts with
- uses: actions-rs/toolchain@v1. The - character is important—that's YAML
syntax for an entry in a sequence. So, step 1 is the first - uses: actions/checkout@
v2 line. Step 2 begins with uses: actions-rs/toolchain@v1, which is the name
of the action we are using. Note that the next line, with:, does not have a dash in front
of it. That's because it's part of the same step, which is a YAML hash with the uses: and
hash: keys. Those fields must line up because YAML is whitespace sensitive. If you're
still confused by YAML, I recommend you do not think about it too much; it's really just
a plain text markup format that works in the way it looks.

In turn, the with key is set to another map with the keys of toolchain, target,
override, and components. They set the toolchain (1.57.0) and target (wasm32-
unknown-unknown) values, and make sure they install the clippy component. Finally,
the override: true flag ensures that this version of Rust is the one in this directory.

With this step, you've added the toolchain you need. However, if you try to run a build
in the workflow, it will still fail because you haven't installed wasm-pack onto the build
machine. You can add that step next, as follows:

```
- uses: actions-rs/install@v0.1
  with:
    crate: wasm-pack
    version: 0.9.1
    use-tool-cache: true
```

You're probably starting to see the pattern. A new step is started with the - character,
and it uses an action. In this case, it is actions-rs/install@v0.1. It's parameters
are the wasm-pack crate, and version 0.9.1. However, we also specify the important
use-tool-cache, which will ensure that if that version of wasm-pack can use a
pre-built binary, it will do so. This shaves several minutes off of your build, so use it
whenever possible.

So, we're ready to build WebAssembly, but there's one more thing to do before we start worrying about building Wasm, and that's running Clippy. When we ran it in *Chapter 9, Testing, Debugging, and Performance*, we did it once manually, but it's important to get this kind of linting into the build so that you catch those kinds of errors early. Typically, I install this kind of check even on my solo projects, because I forget to run it locally. We can add that step like this:

```
- name: Annotate commit with clippy warnings
  uses: actions-rs/clippy-check@v1
  with:
    token: ${{ secrets.GITHUB_TOKEN }}
```

In this case, I've left the `name` field, which was taken straight from the `https://actions-rs.github.io/#clippy-check` documentation. This is because that name will show up on the GitHub Actions UI when it runs, and I might forget what `clippy-check` is. The only parameter it needs is the `token` field, which is set to the magic `${{ secrets.GITHUB_TOKEN }}` field. That field will expand to your actual GitHub API token, which is automatically generated on each workflow run by GitHub Actions. That token is necessary because this action can actually annotate the commit with any warnings that were generated by Clippy, so it needs to be able to write to the repository. The following screenshot shows an example of this where I intentionally introduced a Clippy error:

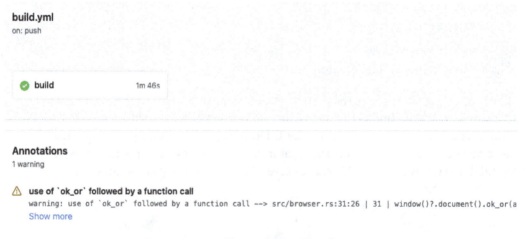

Figure 10.6 – A Clippy error in GitHub Actions

This error also shows up in the commit itself:

```
28    }
29
30    pub fn document() -> Result<Document> {
31  +    window()?.document().ok_or(anyhow!("No Document Found"))

⚠ Check warning on line 31 in src/browser.rs                                        ⊡

   ⊙ GitHub Actions / clippy

   use of `ok_or` followed by a function call

   warning: use of `ok_or` followed by a function call
     --> src/browser.rs:31:26
      |
   31 |     window()?.document().ok_or(anyhow!("No Document Found"))
      |                          ^^^^^^^^^^^^^^^^^^^^^^^^^^^^^^^^^^^ help: try this: `ok_or_else(|| anyhow!("No Document F
      |
      = note: `#[warn(clippy::or_fun_call)]` on by default
      = help: for further information visit https://rust-lang.github.io/rust-clippy/master/index.html#or_fun_call
```

Figure 10.7 – A Clippy error in the commit

This functionality is awesome, but don't introduce Clippy errors to show it off unless you're writing a book; otherwise, it's not safe. Now that we've checked the Rust code for idiomatic errors, it's time to build and run tests. Since this is a Wasm project, for that step, we're going to need Node.js.

Node.js and webpack

The `actions-rs` family of actions is for Rust code, hence the addition of `-rs` at the end of `actions`. So, we're going to need to look elsewhere to install Node.js. Fortunately, installing Node is so common that it's one of the default actions provided by GitHub. We can add another step to set up Node, as shown here:

```
- uses: actions/setup-node@v2
  with:
    node-version: '16.13.0'
```

Any of the actions provided by GitHub can be found in the `actions` repository, and this one is called `setup-node`. In this case, we only need one parameter, `node-version`, which I've set to the **Long-Term Support** (**LTS**) version of Node at the time of writing. This will set Node.js up on the system, but it doesn't run any Node.js tasks. We'll want to run tests and then build the release. This will require three steps, all of which are nice and short and come after the `setup-node` step. They look like the following:

```
- run: npm install
- run: npm test
- run: npm run build
```

Notice how none of these steps have a `uses` key—they just call `run`, which runs the command as written in the shell. Since `Node.js` is installed, you can safely assume `npm` is available, and install, test, and run the build as three more steps in your workflow. This is a great time to commit your workflow and give it a try.

> **Tip**
> Before committing and pushing your code, it can be helpful to run it through a YAML syntax validator. This won't ensure that it's valid for GitHub Actions, but it will at least ensure that it's valid YAML syntax and prevent wasted time pushing simple errors in indentation. `https://onlineyamltools.com/validate-yaml` is an example of a simple online one, and Visual Studio Code has a plugin for it at `https://marketplace.visualstudio.com/items?itemName=redhat.vscode-yaml`.

This build might actually fail at `-run: npm test`, with the following error highlighted:

```
Error: Must specify at least one of `--node`, `--chrome`, `--firefox`, or `--safari`
```

In *Chapter 9, Testing, Debugging, and Performance*, we ran our browser-based tests with the `wasm-pack test --headless --chrome` command. The build script runs `npm test`, which corresponds to the test script in the `package.json` file that was created for us in *Chapter 1, Hello WebAssembly*. If that filename doesn't sound familiar, that's because we haven't spent any time in it. Open it up, and you'll see the test entry, which should look like this:

```
{
  ...
  "scripts": {
```

```
    "build": "rimraf dist pkg && webpack",
    "start": "rimraf dist pkg && webpack-dev-server --open
        -d --host 0.0.0.0",
    "test": "cargo test && wasm-pack test --headless"
  },
...
}
```

In the preceding highlighted code, you can see that it runs `cargo test` and then `wasm-pack test --headless`, but without specifying a browser. There's our build error! You can fix that by adding `--chrome` to the list of parameters passed to `wasm-pack test` and pushing that up to GitHub.

> **Note**
>
> It's possible that this code has been fixed in newer versions of the project skeleton, so you do not see this error. If that's the case, you're already finished—congratulations! It's still useful to understand what tasks are being run under the hood of `npm test`.

At this point, you should have a build that takes about 4 minutes, which is a little longer than I'd like for a small project, but we'll leave optimizing the build to the DevOps team. You've completed the CI step of this section, and now you can move on to the CD part.

Deploying test and production builds

For deployments, we'll use Netlify, which is a cloud computing company that specializes in **single-page applications (SPAs)** like Walk the Dog. It has a generous free tier and a lot of features, such as test deploys with unique URLs, that aren't available with other free solutions such as GitHub Pages. We're going to set up a build that deploys a test version on each push to a branch. Then, when the code has been merged to `main`, it will perform the production build. Production is defined loosely here, as we won't go in great depth into tasks such as getting a custom domain for your app or monitoring for errors, but it's the version of the app that will be publicly available.

In order to deploy from GitHub to Netlify, we'll have to do some wiring so that GitHub has access to push to your Netlify account, and we have a site to push to. So, we're going to use the Netlify CLI to set up a site and prepare it for GitHub pushes. We're not going to use the built-in Netlify-GitHub connection that Netlify provides because it doesn't work with repositories unless you are an administrator on them. In this case, it's also more applicable if you are using other Git providers since the Netlify CLI will work with any of them.

> **Note**
>
> There's an argument to be made that we're not practicing CD here because we won't have our machine completely configured in a tool such as Ansible or Terraform. The Netlify configuration isn't disposable, so it's not CD or DevOps. That's true, but this is not a book about how to configure Netlify in code, so we're not going to concern ourselves with that here. We had to draw a line somewhere.

The first step is to install the CLI itself, which can be installed with `npm install netlify-cli --save` running at the root directory. This will install `netlify-cli` locally, which is in the `node_modules` directory of this project, so it won't pollute your local environment. The `--save` flag automatically adds `netlify-cli` to the list of dependencies in `package.json`.

> **Tip**
>
> If you have trouble running the Netlify CLI, make sure you're using version `16.13.0` of Node.js or higher. There were issues with the earlier versions.

After installing the Netlify CLI, you'll need to call `npm exec netlify login` to log in to your Netlify account. At the time of writing, `npm exec` is the way to ensure you're using the local copy of the `netlify` command, but you could also use `npx` or directly call the copy in `node_modules\.bin`. This will probably change again in the future, so it pays to Google it. The important part is that you probably don't want to install a global version of the `netlify` command unless you know what you're doing.

When you call `npm exec netlify login`, it will take you through the web browser to complete the login process. Then, you'll want to call `npm exec netlify init -- --manual`. The addition of `--` in the middle is important so that `--manual` is passed through to the `netlify` command and not to `npm exec`. You will want to choose **Create & configure a new site**. For your site and team name, you can choose whatever you like, although I've already taken `rust-games-webassembly`. Your build command is `npm run build`, and the directory to deploy is `dist`. You can accept the default settings until the instructions say **Give this Netlify SSH public key access to your repository**. Then, you'll want to copy the provided key and add it to GitHub under your repository's **Settings | Deploy keys** page, as shown in the following screenshot:

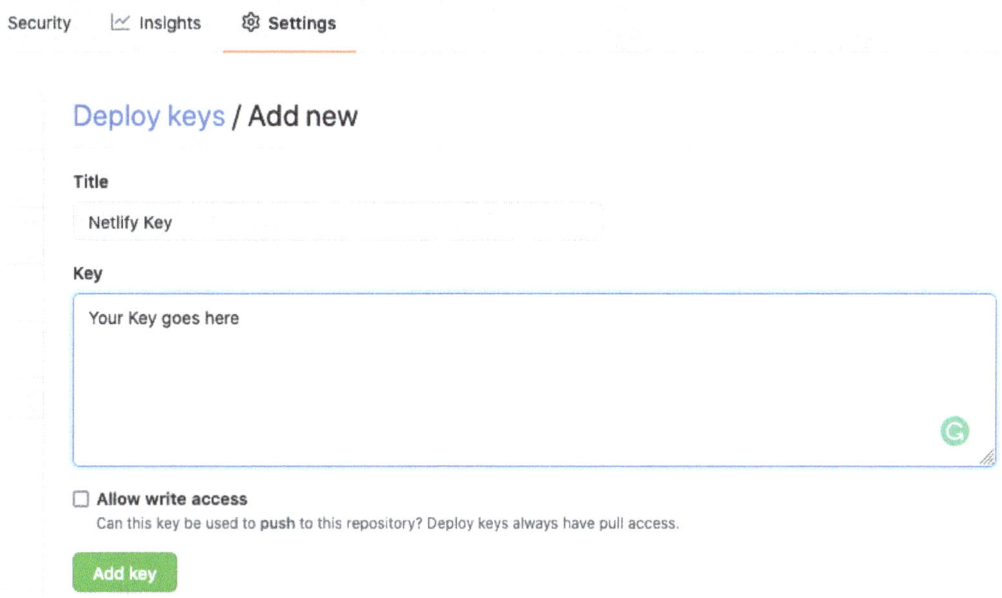

Security   Insights   Settings

Deploy keys / Add new

Title

Netlify Key

Key

Your Key goes here

☐ **Allow write access**
Can this key be used to **push** to this repository? Deploy keys always have pull access.

Add key

Figure 10.8 – Deploying keys

You can accept the default settings, but do not configure the `webhook` setting that is provided. While you can do this, I want to make sure we only push a test build if the build passes, so we'll add this to GitHub Actions instead. That also keeps more of the behavior inside source control. This is because we'll explicitly push to Netlify in a workflow step, whereas configuring through the GitHub GUI means there will be more settings we might forget about if we ever move the repository.

When the command is complete, you should see a message that reads **Success! Netlify CI/CD Configured!**. It will tell you that branches will be automatically deployed when you push to them. Since we didn't set up the webhook, this is incorrect, and there's a little more still to do.

> **Note**
>
> Of course, the CLI might have changed its interface since this book was published. The important takeaway is you want to create the site in Netlify, and do not want to set up a webhook because we'll be using GitHub Actions instead. If the choices have changed, you can look at the official Netlify documentation at `https://docs.netlify.com/cli/get-started/`.

To add the step to deploy to Netlify, we're going to need to add a step to the workflow. That step is as follows:

```
- name: Deploy to Netlify
  uses: nwtgck/actions-netlify@v1.2
  with:
    publish-dir: './dist'
    production-branch: main
    github-token: ${{ secrets.GITHUB_TOKEN }}
    deploy-message: "Deploy from GitHub Actions"
    enable-pull-request-comment: true
    enable-commit-comment: true
    overwrites-pull-request-comment: true
  env:
    NETLIFY_AUTH_TOKEN: ${{ secrets.NETLIFY_AUTH_TOKEN }}
    NETLIFY_SITE_ID: ${{ secrets.NETLIFY_SITE_ID }}
  timeout-minutes: 1
```

We're using the action at `nwtgck/actions-netlify@v1.2` because it has the cool feature of commenting on the commit that does a deployment. There are other actions that use Netlify, and you could also just use the `runs` command after installing the CLI if you so choose. There are many options, and all of this should be considered as an example of one way to set up this workflow and not the actual way to set it up.

The first few flags are somewhat self-explanatory. The build directory is `dist` so that's what we'll publish. The production branch is `main`, and we need `github-token` again so that the action can annotate the commits. The next three flags will enable a PR comment, telling you where the app was deployed to. Put that same comment on the comment, and then overwrite `pull-request-comment` if you deploy the same branch more than once. We've set all of these to true.

The two env fields are probably the most confusing, as they specify a NETLIFY_AUTH_
TOKEN token and the NETLIFY_SITE_ID site ID that you don't have yet. The site ID is
the easier of the two to find, and you can get it through the GUI or the CLI. To get it from
the CLI, run npm exec netlify status in Command Prompt. You should get an
output that looks like this:

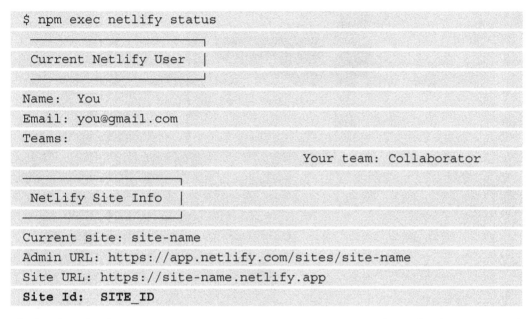

```
$ npm exec netlify status

 ┌──────────────────────────────┐
  Current Netlify User  |
 └──────────────────────────────┘
Name:   You
Email:  you@gmail.com
Teams:
                                    Your team: Collaborator

 ┌──────────────────────────┐
  Netlify Site Info  |
 └──────────────────────────┘
Current site: site-name
Admin URL: https://app.netlify.com/sites/site-name
Site URL: https://site-name.netlify.app
Site Id:   SITE_ID
```

The last line displays your NETLIFY_SITE_ID site ID. You can then take that site ID and
add it to the Secrets section of your GitHub repository, which is located in the **Settings**
tab, with the name of NETLIFY_SITE_ID:

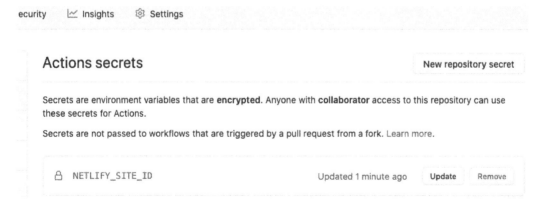

Figure 10.9 – Setting the site ID in GitHub

Also, you'll need a personal access token to access the deployment. That's tricky to find in the Netlify UI, but it's there under **User settings,** which you can find by clicking on your user icon in the upper-right corner of the screen:

Figure 10.10 – User settings

Then, choose **Applications**, not **Security**, and you'll see the **Personal access tokens** section, as shown in the following screenshot:

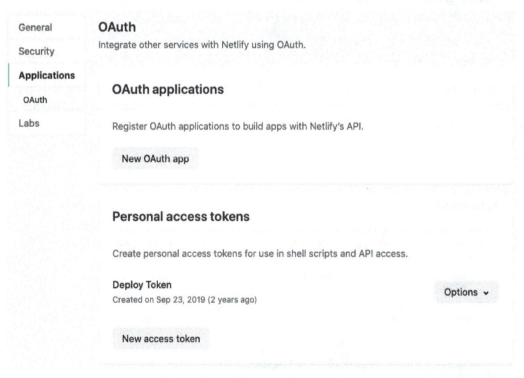

Figure 10.11 – The personal access token

You can see the **New access token** button, so click on it and create a key named `Netlify Deploy` or something similar. Copy that token and add it to the secrets in GitHub, this time, named `NETLIFY_AUTH_TOKEN`:

Actions secrets   New repository secret

Secrets are environment variables that are **encrypted**. Anyone with **collaborator** access to this repository can use these secrets for Actions.

Secrets are not passed to workflows that are triggered by a pull request from a fork. Learn more.

🔒   NETLIFY_AUTH_TOKEN   Updated now   Update   Remove

🔒   NETLIFY_SITE_ID   Updated 18 minutes ago   Update   Remove

Figure 10.12 – Showing two secrets

Once you've added those two keys, you can commit the changes to the workflow, push them up, and you will get an email from the GitHub Actions bot telling you that your app was deployed to a test URL. It's also commented to the commit, which you can see in the following screenshot:

1 comment on commit `17c0db5`

github-actions bot commented on 17c0db5 10 minutes ago   Contributor ☺ ···

🚀 Deployed on https://619f05ac25633b63dae2f99a--rust-games-webassembly.netlify.app

Figure 10.13 – Deployed to test

Alternatively, you can go to the sample repository where you can see the comment at `https://bit.ly/3DR1dS5`. The deploy link in the commit message won't work anymore because it's a test URL, but it did work at one time. That leaves us with one other thing to test. So far, we've been pushing to a branch—at least, you should have been if you paid attention—but if we deploy to the `main` branch, we will get a production deploy. You can get your code over to `main` however you like, merge locally, and push or create a PR. In any case, you just need to push a branch to `main` and you should get a production deployment.

I know I did—you can play Walk the Dog at `https://rust-games-webassembly.netlify.app/`. We shipped!

Summary

Did I mention we shipped? In this chapter, we built a small but functional CI/CD pipeline for the Walk the Dog game. We learned how to create a GitHub Actions workflow and took a tour of how to find actions in the marketplace. Additionally, we started creating both test and production deployments in Netlify. We even get emails when it's done! You could extend this process to do things such as only making the test build on a PR or adding integration tests, and you could use this as a model for other CI/CD pipelines on different systems. This chapter was short, but vital since games must actually ship.

Of course, while the game might be shipped, it's never finished. In the next chapter, we'll discuss some challenges that you can take on to make your version of Walk the Dog superior to the book version. I'm excited to see what you'll do!

11
Further Resources and What's Next?

If you have worked your way through this entire book, reading and writing code in every section, that's fantastic! I don't believe there's a better way to learn the material, and now you have a functioning game. In addition, you probably spent a lot of time debugging when you made mistakes, tweaking when you wanted to have fun, and puzzling over the stranger bits that weren't explained as well as I'd like to think. However, you might still be wondering if you really learned anything, or if you just copied/pasted what I had without understanding it. Don't worry – that's normal, and that is why we're going to do a little bit of a review.

In this chapter, we'll cover the following:

- A challenging review
- Further resources

After this chapter is completed, you'll have validated what you have learned, and I hope to see your games on the web!

Technical requirements

There is a small amount of code in this chapter, found at `https://github.com/PacktPublishing/Game-Development-with-Rust-and-WebAssembly/tree/chapter_11`.

The final version of the game is also available at `https://github.com/PacktPublishing/Game-Development-with-Rust-and-WebAssembly`, and the deployed production version of the game is at `https://rust-games-webassembly.netlify.app/`.

To complete the challenge, you'll need the latest version of the assets at `https://github.com/PacktPublishing/Game-Development-with-Rust-and-WebAssembly/wiki/Assets`.

Check out the following video to see the Code in Action: `https://bit.ly/3JVabRg`

A challenging review

Reviewing code in a book is a strange concept; after all, you can just flip back to the earlier chapters to review what you've learned, so why bother to reiterate that now? At the same time, I've taught a lot of classes, and if there's one thing that's consistent, it's that sometimes smart students sit quietly, listen, nod, and then leave the classroom without understanding anything that you've just said. The only way to get an understanding is to take the knowledge we've practiced so far and build something upon it. Fortunately, we have just the thing.

What happened to the dog?

In *Chapter 2*, *Drawing Sprites*, we did a quick game design session where we described how our little **Red Hat Boy** (**RHB**) would be chasing his dog, who was startled by a cat. Yet, in the proceeding nine chapters, there has been no sign of the dog. Simply put, adding the dog, and a cat, requires very little that you don't already know how to do and would have been redundant. Adding them will be a great way to reinforce what you've done and perhaps learn a new trick or two along the way. To add the dog will require a few steps, intentionally outlined here at a high level:

1. Get the dog sprite sheet into the game: You'll need to take the sprite sheet, found in the `sprite_sheets` folder in the assets with the name `dog_sheet`. That's the dog in his running animation, ready to be put into place. Look at *Chapter 2*, *Drawing Sprites*, to remind yourself how that works.

Add a dog `struct`: There will need to be a dog `struct` in the game as one of the many game objects. It will look similar to the `RedHatBoy` object, which, as you've probably guessed, means you are likely to need a state machine, as we covered in *Chapter 4, Managing Animations with State Machines*. What would you use a state machine for? To make sure that the dog goes to the right at the start of the game, only to have him turn around and run back to RHB when RHB crashes. You would have states for running right and running left. The dog should also hold still right at the beginning, making sure to only take off after a moment, after which RHB gives chase.

2. Extend the `WalkTheDogStateMachine`: For the dog to hold still, and for RHP to ignore user commands, you're going to need to extend `WalkTheDogStateMachine` beyond the `Ready` state. We covered all of this in *Chapter 8, Adding a UI*.

    Of course, that's a simple way to add the dog, but this being a video game, you're only limited by your imagination. Probably the simplest thing to do would be to have the dog run off screen, and then run back after RHB falls over. You can also keep the dog on screen and have him safely navigate the platforms the same way that the player attempts to. That will mean a few more changes.

3. Add hints to the endless runner: In *Chapter 6, Creating an Endless Runner*, we created segments of the game based on where the player was and a random value. Each segment can also have "hints" for the dog, so it knows when to jump to get around the various obstacles.

4. Make sure the dog barks: As a dog owner, there's one thing I know about them – they are not *silent*. Our dog should make noise, such as barking, using the same technology we covered in *Chapter 7, Sound Effects and Music*. You can also add some running sound effects, as well as a crash when the user fails to get by a platform or hits a rock.

5. Keeping score: The game doesn't really keep score, and it could. It uses a time-based model, racking up points the longer the player stays alive and adding bonuses every time the player completes a jump on a platform or slides under a box. There are a ton of choices. You'll keep that score in the `Game` object we initially implemented in *Chapter 3, Creating a Game Loop*, and display it using the same technology we used in *Chapter 8, Adding a UI*.

6. Using slide: The tiles sprite sheet has a lot more graphics than just the little islands and the rock we've used so far. We've also got a slide animation, but we don't have anything short enough to slide under. Using the techniques from *Chapter 6, Creating an Endless Runner*, set up a segment that the player can slide under.

It's a cliché, but the limits are really your imagination. Years ago, I taught a workshop on HTML5 game development where I provided the students with an *Asteroids* clone to start with. One of them returned the next week with a Mario-like platformer!

> **Tip**
> Remember that each chapter of this book is reachable from a Git tag at the repository `https://github.com/PacktPublishing/Game-Development-with-Rust-and-WebAssembly`. In addition, the main branch contains the entire game, including my solutions to these challenges as they are completed. If you purchased this book early enough, you can even see me work on them live at `www.twitch.tv/paytonrules`.

Further resources

After working through this game and completing some of the challenges that I just mentioned, maybe you want to go even bigger with your next game. I hope you do. You can add particle effects, explosions, or an online scorekeeping system. You can also use this framework as the start of a completely original game. You can also decide to use this game as an introduction and start a completely new game of your own using a completely different framework. This section is meant to show you just a few of the options available to you now if you want to keep making games, especially with Rust and WebAssembly.

Using JavaScript libraries

This entire game has been written using Rust as our language of choice, effectively discarding the entire JavaScript ecosystem. That's been a deliberate choice, but it's not the only one. We could also have called into a Rust Wasm library from an existing JavaScript framework or could have used `wasm-bindgen` to enable calling out to a JavaScript library or framework from Rust code. The first is more practical, and a great way to introduce Rust into an existing JavaScript project. The second one is more fun, so naturally, we'll be taking a brief peek at that one, with an example written using PixiJS.

PixiJS

PixiJS (`https://pixijs.com/`) is a popular and productive JavaScript framework for making games and visualizations in JavaScript. It has a Canvas and WebGL-backed renderer, and it's a great way to get high-performance 2D graphics without writing WebGL shaders yourself. It supports a ton of cool features and is a lot faster than using the Canvas as we did in our game. It has screenshots like this:

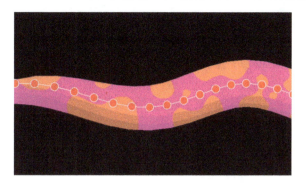

Figure 11.1 – A textured mesh (https://bit.ly/3JkhbXw)

It is also a lot more complicated than our engine, which is one reason why this book does not use it, but it's great to try on your *next* game. To use JavaScript libraries from Rust code, you need to import functions using the wasm-bindgen library, as follows:

```
#[derive(Serialize, Deserialize)]
struct Options {
    width: f32,
    height: f32,
}


#[wasm_bindgen]
extern "C" {
    type Application;
    type Container;

    #[wasm_bindgen(method, js_name = "addChild")]
    fn add_child(this: &Container, child: &Sprite);

    #[wasm_bindgen(constructor, js_namespace = PIXI)]
    fn new(dimens: &JsValue) -> Application;

    #[wasm_bindgen(method, getter)]
    fn view(this: &Application) -> HtmlCanvasElement;

    #[wasm_bindgen(method, getter)]
    fn stage(this: &Application) -> Container;
```

```rust
    type Sprite;

    #[wasm_bindgen(static_method_of = Sprite, js_namespace
        = PIXI)]
    fn from(name: &JsValue) -> Sprite;
}

// This is like the `main` function, except for JavaScript.
#[wasm_bindgen(start)]
pub fn main_js() -> Result<(), JsValue> {
    let app = Application::new(
        &JsValue::from_serde(&Options {
            width: 640.0,
            height: 360.0,
        })
        .unwrap(),
    );

    let body =
        browser::document().unwrap().body().unwrap();
    body.append_child(&app.view()).unwrap();

    let sprite =
        Sprite::from(&JsValue::from_str("Stone.png"));

    app.stage().add_child(&sprite);

    console_error_panic_hook::set_once();

    Ok(())
}
```

I've hidden the use declarations, but this is a version of lib.rs from our game that just uses PixiJS to render a static screen. It's not much fun yet, but it's enough to demonstrate how, using the wasm_bindgen macro and the extern "C" struct, you can import any JavaScript functions into your Rust library that you may want to use. This allows you to use arbitrary JavaScript code in your Rust program, with a little bit of glue code to wire the parts together. In fact, this is exactly how web_sys, which we've been using all over the place, works.

In order to use all that Pixi code, you'll need to add a reference to the pixi.js JavaScript library, and a quick and dirty way to do this is to add the following to index.html:

```html
<!DOCTYPE html>
<html>
<head>
  <meta charset="UTF-8">
  <title>My Rust + Webpack project!</title>
  <link rel="stylesheet" href="styles.css" type="text/css"
      media=          "screen">
  <link rel="preload" as="image" href="Button.svg">
  <link rel="preload" as="font" href=
  "kenney_future_narrow-webfont.woff2">
  <script src="https://pixijs.download/release/pixi.js">
  </script>
</head>
...
```

In a professional deployment environment, you'd probably want to use WebPack to bundle the JavaScript with your own source code, but this works for now. I've also removed our canvas element from the HTML because Pixi provides its own.

In the Rust code, I was able to import the `PIXI.Application`, `PIXI.Container`, and `PIXI.Sprite` types from `pixi.js`, and I've also pulled in quite a few functions associated with them. This allowed me to use them in `main_js`, just like native Rust code. The example here is not professional, using `unwrap` all over the place, but it successfully creates a PixiJS application and then creates `Sprite` from a file we already had in our game. Then, it adds it to `stage`, which is a PixiJS concept that you can think of as the canvas. This code leads to a screen that looks like this:

Figure 11.2 – A rock

Okay, it doesn't look like much, but the point is that you can use PixiJS in a Rust project by declaring the types you need using `wasm-bindgen`. We won't be covering all that here, but the docs for `wasm-bindgen` are extremely thorough at `https://rustwasm.github.io/wasm-bindgen/reference/attributes/index.html`.

More importantly, maybe you don't like PixiJS, and you want to use **PhaserJS**; the same principle applies! You can use any of the great frameworks available to JavaScript programmers for game development, such as **Three.JS** and **Babylon3D**, provided you can include them in your WebAssembly project. But what if you don't want to use JavaScript at all but still want to run on the web?

Macroquad

Macroquad (`https://macroquad.rs/`) is one of many game development libraries written in Rust. The authors refer to it as a "**game library**", which is a way of saying it's not as fully featured as an entire framework, but it's more featured than just writing to the HTML Canvas element, as we did in our game. It supports WebAssembly out of the box, without writing any JavaScript. An example of the code in Macroquad is shown here:

```
use macroquad::prelude::*;


#[macroquad::main("BasicShapes")]
```

```
async fn main() {
    loop {
        clear_background(RED);

        draw_line(40.0, 40.0, 100.0, 200.0, 15.0, BLUE);
        draw_rectangle(screen_width() / 2.0 - 60.0, 100.0,
            120.0, 60.0, GREEN);
        draw_circle(screen_width() - 30.0, screen_height()
            - 30.0, 15.0, YELLOW);
        draw_text("HELLO", 20.0, 20.0, 20.0, DARKGRAY);

        next_frame().await
    }
}
```

This very simple example will compile and run on the web simply by specifying the target with `cargo build --target wasm32-unknown-unknown` – no JavaScript, no problem. Macroquad is great, but it's not really a full engine. So, what if you want that experience?

Bevy

Another choice with more features is **Bevy** (`https://bevyengine.org/`), which has been extremely popular since its initial announcement and supports WebAssembly. Its "Hello World" is very different from the Macroquad version and resembles the following:

```
use bevy::prelude::*;

fn main() {
    App::new().add_system(hello_world_system).run();
}

fn hello_world_system() {
    println!("hello world");
}
```

The most unique part of this system is the `add_system` function, which allows you to add "systems" to the Bevy engine. Bevy uses a modern Entity Component System for its development, which is meant to aid in structuring your program as well as performance. It's gaining popularity extremely rapidly and moving faster than its documentation can keep up with. Currently, if you're looking to learn how to use Bevy for 2D and 3D games, your best bet is to get involved with the community here: `https://bevyengine.org/community/`. If you do, you'll be rewarded, as Bevy is a very advanced engine, but it doesn't have an editor such as Unity3D or Unreal. If you're looking for that, fortunately, you have an excellent option.

Godot

My first experiences with game development in Rust were using the Godot game engine (`https://godotengine.org`). Godot is a truly free and open source engine that's popular with hobbyists and professional game developers alike. It comes with its own built-in language, GDScript, out of the box but is also able to use Rust through its GDNative wrappers. Originally devised to allow the use of C and C++, GDNative works extremely well with Rust. It has a thriving community of its own, and you can download it here: `https://godot-rust.github.io`.

Using Godot will mean getting a fully featured 2D and 3D engine that's capable of competing with Unity3D at its best. It's possible that the entire time you were reading this book, you wanted to see a proper commercial game engine such as this:

Figure 11.3 – A Godot game engine

If so, Godot is the one for you. To see an example Godot program written in Rust, you can see the one I wrote at `https://github.com/paytonrules/Aircombat`.

Summary

The website `https://arewegameyet.rs` asks the question, "*Is Rust ready for game development?*" and answers with, "*Almost.*" Respectfully, because it's a really cool site, I disagree. We have all of the tools that JavaScript developers had a few years ago, with all the advantages of an excellent type system and Wasm. We have far more tools than developers have had for most of game development history, and while we may not have our Unity or Unreal yet, we have everything we need to *build our own*. So, go out there, build your own games, extend the engine, and have fun! I hope I hear from you with far better games than this one. If you need help, want to show off your games, or just want to hang out with like-minded people, you can find me on the Rustacean Station Discord at `https://discord.gg/cHc3Gyc`. You can always find me as `@paytonrules` on Twitter, and I'd be really excited to hear from you.

Index

X

Y

Hi!

I'm Eric Smith, author of *Game Development with Rust and WebAssembly*, I really hope you enjoyed reading this book and found it useful for increasing your productivity and efficiency in Rust game development.

It would really help us (and other potential readers!) if you could leave a review on Amazon sharing your thoughts on the book.

Go to the link below or scan the QR code to leave your review:

`https://packt.link/r/1801070970`

Your review will help us to understand what's worked well in this book, and what could be improved upon for future editions, so it really is appreciated.

Best wishes,

Eric Smith

Packt.com

Subscribe to our online digital library for full access to over 7,000 books and videos, as well as industry leading tools to help you plan your personal development and advance your career. For more information, please visit our website.

Why subscribe?

- Spend less time learning and more time coding with practical eBooks and Videos from over 4,000 industry professionals

- Improve your learning with Skill Plans built especially for you

- Get a free eBook or video every month

- Fully searchable for easy access to vital information

- Copy and paste, print, and bookmark content

Did you know that Packt offers eBook versions of every book published, with PDF and ePub files available? You can upgrade to the eBook version at packt.com and as a print book customer, you are entitled to a discount on the eBook copy. Get in touch with us at customercare@packtpub.com for more details.

At www.packt.com, you can also read a collection of free technical articles, sign up for a range of free newsletters, and receive exclusive discounts and offers on Packt books and eBooks.

Other Books You May Enjoy

If you enjoyed this book, you may be interested in these other books by Packt:

Creative Projects for Rust Programmers

Carlo Milanesi

ISBN: 978-1-78934-622-0

Access TOML, JSON, and XML files and SQLite, PostgreSQL, and Redis databases.

- Develop a RESTful web service using JSON payloads.
- Create a web application using HTML templates and JavaScript and a frontend web application or web game using WebAssembly.
- Build desktop 2D games.

Rust Web Programming

Maxwell Flitton

ISBN: 978-1-80056-081-9

- Structure scalable web apps in Rust in Rocket, Actix Web, and Warp.
- Apply data persistence for your web apps using PostgreSQL.
- Build login, JWT, and config modules for your web apps.
- Serve HTML, CSS, and JavaScript from the Actix Web server.
- Build unit tests and functional API tests in Postman and Newman.

Packt is searching for authors like you

If you're interested in becoming an author for Packt, please visit authors. packtpub.com and apply today. We have worked with thousands of developers and tech professionals, just like you, to help them share their insight with the global tech community. You can make a general application, apply for a specific hot topic that we are recruiting an author for, or submit your own idea.

www.ingramcontent.com/pod-product-compliance
Lightning Source LLC
Chambersburg PA
CBHW081456050326
40690CB00015B/2819